ICT activities
for mathematics

David Benjamin
Justin Dodd

Websites

Links to appropriate websites are given in some worksheets. Although these were up to date at the time of writing, it is ESSENTIAL for teachers to preview these sites before using them with pupils. This will ensure that the web address (URL) is still accurate and the content is suitable for your needs.

We suggest that you bookmark useful sites and consider enabling pupils to access them through the school intranet. We are bringing this to your attention as we are aware of legitimate sites being appropriated illegally by people wanting to distribute unsuitable or offensive material. We strongly advise you to purchase suitable screening software so that pupils are protected from unsuitable sites and their material.

If you do find that the links given no longer work, or the content is now unsuitable, please let us know. Details of changes will be posted on our website. You can also check our website for any changes that we know of.

Software

We assume that you will have access to the following software packages:
Microsoft Internet Explorer 5 (or higher)
Microsoft Excel 97 (or higher).

In addition, to make full use of the activity sheets in this pack, pupils will also need access to full working versions of the following software packages:
The Geometer's Sketchpad
Omnigraph
MSWLogo.

Heinemann Educational Publishers
Halley Court, Jordan Hill, Oxford OX2 8EJ
a division of Reed Educational & Professional Publishing Ltd
Heinemann is a registered trademark of Reed Educational & Professional Publishing Ltd

OXFORD MELBOURNE AUCKLAND
JOHANNESBURG BLANTYRE GABORONE
PORTSMOUTH NH (USA) CHICAGO

First published 2002
ISBN 0 435 41834 3

06 05 04 03 02
10 9 8 7 5 4 3 2 1

Designed and typeset by Wendi Watson
Illustrated by Cedric Knight
Cover design by Miller, Craig and Cocking
Printed and bound in Great Britain by Athenaeum Press Ltd

Acknowledgements
The publishers would like to thank Richard Knights for designing and producing the CD-ROM
for this pack.

Publishing team
Editorial
Sue Bennett
Jim Newall
Nick Sample
Harry Smith

Design
Phil Richards
Colette Jacquelin

Production
David Lawrence
Jason Wyatt

Author team
David Benjamin
Justin Dodd
Deborah Stanley

Tel: 01865 888058 www.heinemann.co.uk

ICT activities for mathematics

1 Introduction

This pack provides activity sheets for use in Mathematics at Key Stages 3 and 4 as well as detailed support for the teacher, including guidance on using ICT in general.

The activity sheets in the pack use four main software packages:
- Microsoft Excel
- Omnigraph
- MSWLogo
- The Geometer's Sketchpad.

A copy of MSWLogo is included on the CD-ROM along with 30-day demonstration copies of Omnigraph and The Geometer's Sketchpad. Details of how to obtain the software, and suggestions for where to look for other software are given in the section Software available.

The pack is split into the following sections:
- **Using ICT in mathematics**: provides guidance on computers and software, as well as an introduction to using the packages that are used in the worksheets.
- **Teaching notes**: provides guidance on selecting appropriate worksheets for you to use with your pupils, as well as notes on specific worksheets and answers.
- **Activity sheets**: 56 Activity sheets for Key Stages 3 and 4.
- **CD-ROM**: contains copies of all Activity sheets in portable document file (PDF/Acrobat) format.

Using ICT in mathematics

This section contains detailed guidance for teachers on many aspects of using ICT in mathematics. It is designed to be a reference that teachers can dip into if there is something they may be unfamiliar with, or to look for further information.

- **Availability of computers and software available**. Describes each of the software packages used in the activities and gives details of how you can obtain the full version of the software. Also contains guidance on finding further software to use with your pupils.
- **Guides to using Excel/Omnigraph/MSWLogo/The Geometer's Sketchpad**. These take the new user through the main functions of the software to enable a new user to get started. All the functions that are used in the activity sheets are covered here, along with guidance on some of the common problems and limitations of the packages.
- **Searching the internet**. Provides guidance for the new user on how to search for particular items on the internet and a list of some useful websites.
- **Using an intranet**. Shows how an intranet can be used in a school to disseminate information and provide a useful resource.
- **Creating high-quality worksheets**. This section shows how professional looking worksheets can be created using features such as Equation Editor in Microsoft Word and a drawing package.
- **Projecting your software onto a screen**. Introduces the use of data projectors and electronic whiteboards that can be used as an interactive visual display for the whole class.

Teaching notes

This section contains notes and tips on using the activity sheets.

- **General use of the ICT Room.** Some tips to help you provide a better lesson in the ICT room.
- **Using the worksheets and CD-ROM.** Guidance on structuring a lesson using an activity sheet and using the CD-ROM, including networking the CD-ROM.
- **Choosing an activity sheet.** This section contains a description of each activity sheet, along with the package used, any files that are needed from the CD-ROM and any assumed knowledge that the pupils need. If pupils' assumed knowledge is weak, suggestions are given for activity sheets that the pupils can do to improve their assumed knowledge. There is also a National Curriculum reference for each activity sheet.
- **Teaching notes.** This contains notes on each activity sheet, along with answers, where appropriate. The teaching notes also contain detailed sheet by sheet notes which list any common problems and errors that pupils (and teachers!) make.

Worksheets

These are grouped under four headings

- Excel
- Ma2: Number and algebra
- Ma3: Shape, space and measure
- Ma4: Handling data

Unfamiliar with ICT or a particular package?

For those teachers who are not familiar with the software packages used in this pack, **Using ICT in Mathematics** contains an overview of each software package along with some advice on getting started and performing standard tasks in each software package. The best way to become proficient with the software is to teach yourself (using the overviews) and then use the software regularly. Practice is the key to success! When you do not use a particular software package for a while it is easy to forget what to do.

Try using the activity sheets yourself beforehand so that you are familiar with what the pupils will be doing.

There is also guidance on searching the internet, using an intranet, producing high-quality worksheets and how to project software on to a screen so that you can demonstrate how to do things to the whole class.

What activity sheet should I use with my class?

In the teaching notes, there are summaries of each activity sheet, along with details of the package used and any files that are required. National Curriculum levels are also given along with any skills that the pupils need to be able to carry out the activities on the activity sheet. Suggestions are also given for activity sheets that you can use if your pupils are weak in any of these skills.

How to find your way around the pack

If you are unfamiliar with any aspect of ICT, use the following table
to help identify what you need to do.

I want to know how to use **Excel**	Read the guide on using Excel and practice using Excel.	Go to pp 6–13
I want to know how to use **Omnigraph**	Read the guide on using Omnigraph and practice using it.	Go to pp 14–16
I want to know how to use **Logo**	Read the guide on using Logo and practice using it.	Go to pp 17–20
I want to know how to use **The Geometer's Sketchpad**	Read the guide on using The Geometer's Sketchpad and practice using it.	Go to pp 21–23
I'm looking for some suggestions for suitable **websites**	See 'Website information'.	Go to pp 27–28
I want to know how to conduct a **search** on the **internet**	See 'Searching the internet'.	Go to pp 24–26
I would like to set up an **intranet**	See 'Using an intranet'.	Go to pp 29–31
Which **activity sheet** should I use for my class?	See 'Selecting an activity sheet'.	Go to pp 50–58
How can I **create** better looking **worksheets**?	See 'Creating high-quality worksheets'.	Go to pp 32–43
How can I **project software** onto a screen for a whole class demonstration?	See 'Projecting your software onto a screen'.	Go to pp 44–46
Are there any **pitfalls** or **common problems** to look out for when using the software?	In the teaching notes for each worksheet, any pitfalls and common problems are listed.	Go to pp 63–89
How do I use the **CD-ROM**?	See 'Using the CD-ROM'.	Go to p 49

2 | Using ICT in mathematics

Information and communications technology (ICT) can be used to extend pupils' learning in mathematics. The use of ICT can help to raise pupils' standards of achievement and enhance their enjoyment of the subject.

The use of ICT is part of good classroom practice. All pupils can use ICT, individually, in groups or as a whole class. ICT can be used by students for research, interactive activities, investigative work, homework, printing external examination questions, examiners' reports and solutions, and for the presentation of coursework. As well as using ICT to help in the classroom, you can use ICT to help with lesson planning, research, record keeping and analysis, report writing and professional development.

Professional development can be linked to Performance Management, introduced in Autumn 2000 to replace Teacher Appraisal. Using ICT in its various forms and keeping records of pupils' progress are important parts of these two initiatives, and will provide data for the Threshold Assessment Application Form.

Availability of computers

Your school should provide the necessary facilities for you to use ICT with a whole class and for a satisfactory number of sessions. At present, not all schools provide this, but resources are improving. A lot can still be done with a few departmental machines, but this inevitably makes classroom management more difficult.

Software available

- **Spreadsheets**
  Most schools will have a spreadsheet program and this will usually be a version of Microsoft Excel. The latest version is Excel2000. This pack provides a number of activities for spreadsheet use ranging from 'animated' probability modelling to simple formulae creation. Excel is a very powerful tool and takes students beyond the limits of a calculator.

- **Graphing packages**
  These are a must for Mathematics departments. The activity sheets in this pack are designed for use with Omnigraph. This is an excellent piece of software *and* easy to use. It enables students to investigate linear, quadratic, cubic, reciprocal and gradient functions, to name a few! You can study function and geometric transformations and trigonometric identities.

  A 30-day evaluation version of Omnigraph can be downloaded from http://www.spasoft.co.uk.

  Omnigraph is available from Software Production Associates, PO Box 59, Tewkesbury GL20 6AB; tel. : 01684 833700; http://www.spasoft.co.uk.

- **Dynamic geometry software**
  The activity sheets included here are based on The Geometer's Sketchpad. It is perhaps the most difficult of the software included, although 30 minutes practise should be enough! However, it is very

powerful and an excellent way to study the properties of geometric shapes. A student will always remember that the angle in a semicircle is 90° after they have dragged the point H below around the circumference.

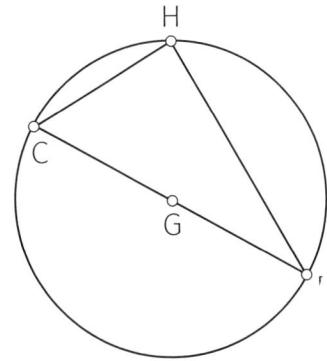

The Geometer's Sketchpad is available from Virtual Image, 184 Reddish Road, South Reddish, Stockport SK5 7HS; tel. : 0161 480 1915; http://www.virtualimage.co.uk/.

- **Control technology**
  A great way to get students to understand polygons is to ask them to produce the instructions necessary to draw them. MSWLogo is a program designed for this type of activity. It is easy to use and good fun. Super Logo and WinLogo are similar programs for this type of activity.

MSWLogo can be downloaded for free from:
http://www.softronix.com/logo.html.

- **National Curriculum topics**
  Virtual Image (details above) and MicroSmile (Smile Mathematics, Isaac Newton Centre, 108A Lancaster Road, London W11 1QS; tel. : 020 7598 4841; http://www.smilemathematics.co.uk) produce software designed to allow students interactive practice and to investigate, respectively, National curriculum topics.

There is a lot of Mathematics software available and these are listed in, for example, the AVP and REM catalogues, which should be delivered to your school on a regular basis.

The best way to evaluate any software is to try it out yourself. You should consider the following when looking at software:
- ease of use
- cost
- how it will fit your schemes of work
- whether or not your students will benefit mathematically from its use.
It is time consuming to order and evaluate software but your own department's views are the best.

You can also ask your LEA Mathematics advisors for advice, visit other schools and visit the BETT exhibitions in London and Birmingham. The NGfL website offers some software reviews as well as a discussion group where you can ask colleagues, both locally and in other Authorities and countries, to recommend software.

CD-ROMs with external examination papers on them.
Edexcel GCSE Exambank 2000 (0 435 53254 5)
Edexcel A-Level Exambank 2001 (0 435 51817 8)

A guide to using Microsoft Excel

There are a lot of large textbooks on the market explaining how to use the spreadsheet package Excel and each new version (the current version is Excel 2000) brings a new set of books! The following pages are a 'quick-start' guide to Excel, showing you how to get started and some tips and hints to help you get the most from the software.

Getting started ...

When the program loads you will be presented with a worksheet similar to the one below. This is *sheet 1*. You can work on the other sheets as well and the whole set of sheets is called a *workbook*.

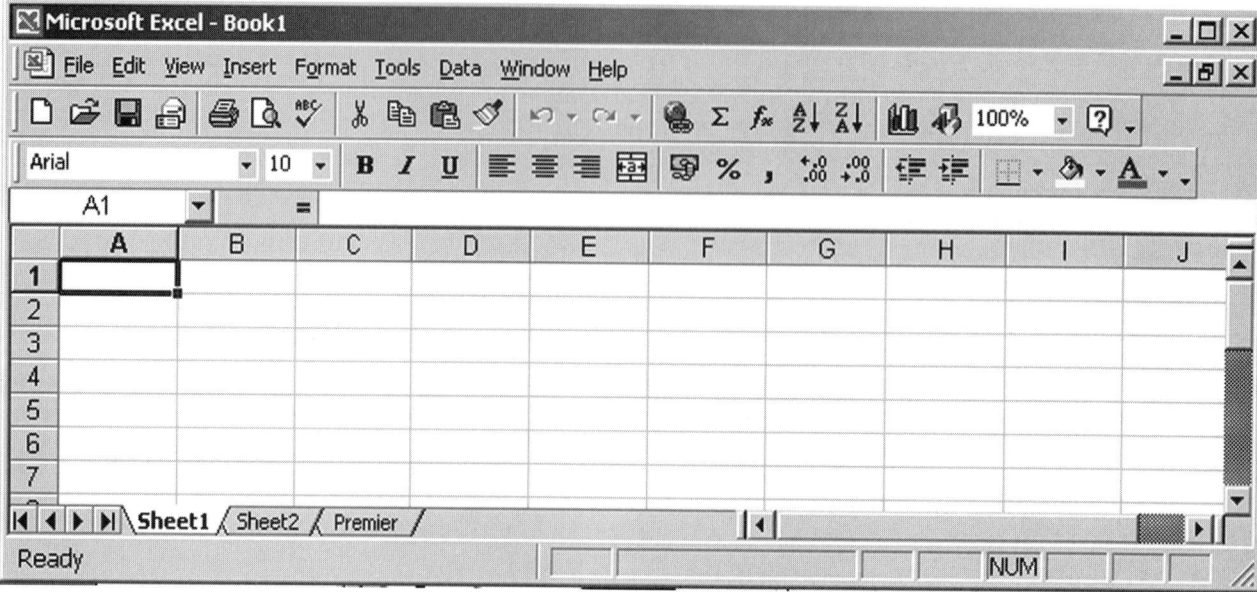

The diagram above is from an Excel97 workbook and contains approximately:

- 256 columns, labelled: A, B, C, ...AA, AB, ...
  AZ, IA, IB, ...IZ
- 65,536 rows
- 16,777,216 cells

You can think of Excel as the ultimate graphical calculator with, for example:

- the power to display more than 65,000 decimal places of a decimal fraction – see Activity NA4 on page 102.
- the power to analyse a variety of statistical data – see Activity HD9 on page 160.
- the power to follow trends so that you can predict future Olympic records – see Activity HD5 on page 155.
- the power to look at the infinite – see Activity NA15 on page 116

In your worksheet:

- The formula bar can be added and removed by clicking <u>V</u>iew on the menu bar and then selecting or deselecting it. The formula bar displays the formula you are entering into a cell.
- The toolbars can be added or removed by clicking <u>V</u>iew and then <u>T</u>oolbars.
- The most useful toolbars are those shown on page 6 and the drawing toolbar which allows you to add textboxes and shapes to your worksheet.
- Use the Ctrl key on your keyboard to highlight cells, rows, or columns, which are *not* adjacent to each other. This is useful when wishing to display your data as a chart.

	A	B	C	D	E
1	Club	Home Wins	Home Draws	Home Defeats	Home Goals Scored
2	Nottingham F.	18	2	3	52
3	Middlesborough	17	4	2	51
4	Sunderland	14	7	2	49
5	Charlton	17	5	1	48
6	Ipswich	14	5	4	47

Entering data

To enter data into a worksheet, click on a cell to highlight it and enter the data into the cell. You can enter a number of different types of data into a cell, the most common being numbers and formulae. Numbers are just entered as they are written – use a full stop for a decimal point and prefix the number with a minus sign if it is negative. Fractions are simply entered using the division sign, /. (See the notes on formatting cells if the data doesn't display how you intended.)

You enter formulae in the same way as numbers, but they must be prefixed with an equals sign to ensure they display the result in the highlighted cell. An example of a formula is =A2/2. Some useful entries are:

- Indices – Enter these by using the caret symbol, ^
  (eg, enter 4^3 as = 4^3).
- Square roots – type in SQRT instead of the root sign (eg, $\sqrt{2}$ is entered as =SQRT(2)).
- Pi – in Excel you enter pi as PI().
- Trigonometric functions – Excel works in radians only, so you need to express angles in radians. For example, to display cos 90°, enter
  =COS(90*PI()/180).

You can also enter formulae to perform operations on data held in other cells. To do this, type in the name of the cell which contains the data you want to perform the operation on. The table below shows how you can use some of the common formulae to do this:

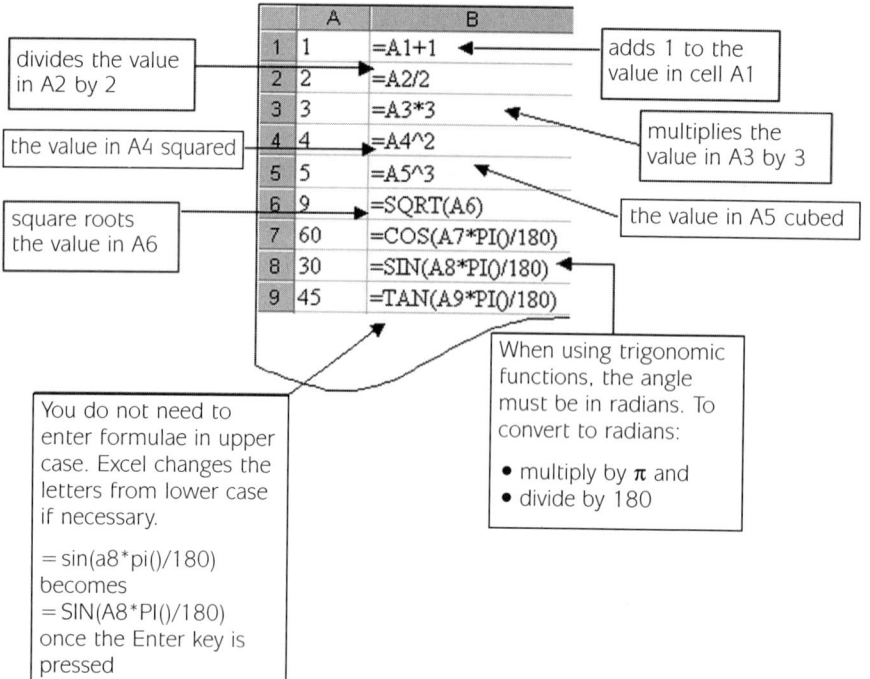

	A	B
1	1	=A1+1
2	2	=A2/2
3	3	=A3*3
4	4	=A4^2
5	5	=A5^3
6	9	=SQRT(A6)
7	60	=COS(A7*PI()/180)
8	30	=SIN(A8*PI()/180)
9	45	=TAN(A9*PI()/180)

divides the value in A2 by 2

adds 1 to the value in cell A1

the value in A4 squared

multiplies the value in A3 by 3

square roots the value in A6

the value in A5 cubed

You do not need to enter formulae in upper case. Excel changes the letters from lower case if necessary.

= sin(a8*pi()/180)
becomes
= SIN(A8*PI()/180)
once the Enter key is pressed

When using trigonomic functions, the angle must be in radians. To convert to radians:

• multiply by π and
• divide by 180

You can also perform operations on groups of cells to help with statistical calculations:

- =sum(A1:A6) sums all the data from cell A1 to cell A6. Alternatively, you can highlight cells A1 to A6 and then click the AutoSum button on the Standard toolbar. **Σ**

- =average(A1:A6) gives the *mean* average of the data in cells A1 to A6
- =median(A1:A6) gives the *median* average of the data in cells A1 to A6
- =mode(A1:A6) gives the *modal* average of the data in cells A1 to A6
- =quartile(A1:A6,1) gives the *lower* quartile of the data in cells A1 to A6
- =quartile(A1:A6,2) is another formula for the *median* of the data in cells A1 to A6
- =quartile(A1:A6,3) gives the *upper* quartile of the data in cells A1 to A6.
- =stdev(A1:A6) gives the *standard deviation* of the data in cells A1 to A6

You can find many more functions by clicking <u>I</u>nsert then <u>F</u>unction to display the Paste Function dialogue box:

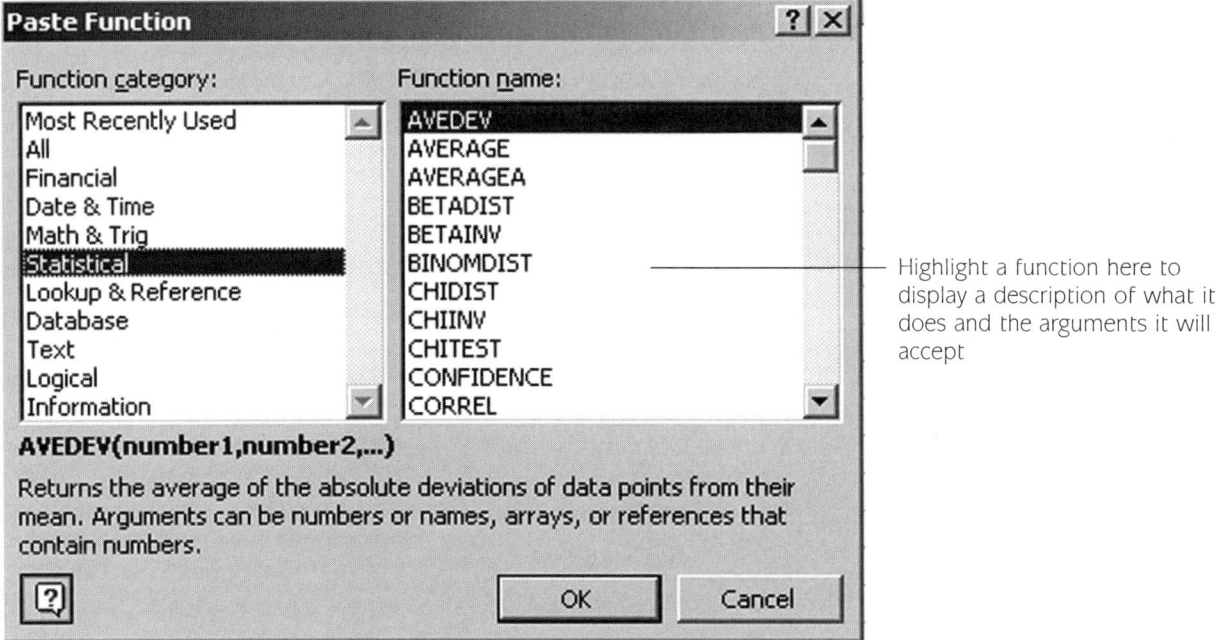

Highlight a function here to display a description of what it does and the arguments it will accept

Copying Formulae

A quick way to copy a formula down, up or across is to highlight the cell with the formula in and drag the black box in the right-hand bottom corner of the cell:

Drag this square down to row 6…

to produce this sequence

…from these formulae.

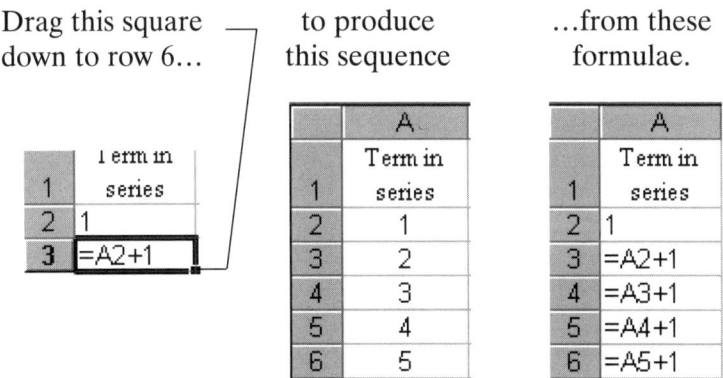

	A
	Term in
1	series
2	1
3	2
4	3
5	4
6	5

	A
	Term in
1	series
2	1
3	=A2+1
4	=A3+1
5	=A4+1
6	=A5+1

If you don't want A2 changed to A3, A4, etc. when you copy the formula you should use the $ sign. Suppose you want to model the changes to profit and loss of a small shop by looking at percentage increase and decrease. As you change cell C2 the selling prices in column B change automatically:

Figure 1

	A	B	C
1	Cost Price	Selling Price	Percentage change
2	1.25	=A2*C2	0.88
3	2.36	=A3*C2	
4	3.25	=A4*C2	
5	6.87	=A5*C2	

Figure 2

	A	B	C
1	Cost Price	Selling Price	Percentage change
2	£ 1.25	£1.25	1.00
3	£ 2.36	£2.36	
4	£ 3.25	£3.25	
5	£ 6.87	£6.87	

Figure 3

	A	B	C
1	Cost Price	Selling Price	Percentage change
2	£ 1.25	£1.38	1.10
3	£ 2.36	£2.60	
4	£ 3.25	£3.58	
5	£ 6.87	£7.56	

Figure 4

	A	B	C
1	Cost Price	Selling Price	Percentage change
2	£ 1.25	£1.10	0.88
3	£ 2.36	£2.08	
4	£ 3.25	£2.86	
5	£ 6.87	£6.05	

Figure 1 shows the formulae for a 12% decrease. Note how C2 has not become C3, C4 and C5 because of the $ sign in front of both the 'C' and the '2'.

Figure 2 shows the Selling Price and the Cost Price before any changes.

Figure 3 shows a 10% increase.

Figure 4 shows a 12% decrease.

Displaying Formulae

If you need to display the formulae in your cells instead of the result, click on <u>T</u>ools and then <u>O</u>ptions..., then click the View tab and add a tick next to Fo<u>r</u>mulas:

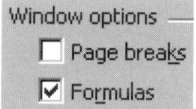

You can easily reverse this process to display the results again later.

Formatting cells

You can format cells to display data in a variety of ways. Highlight the cells you wish to format and then click F<u>o</u>rmat and C<u>e</u>lls on the menu bar to produce the following dialogue box:

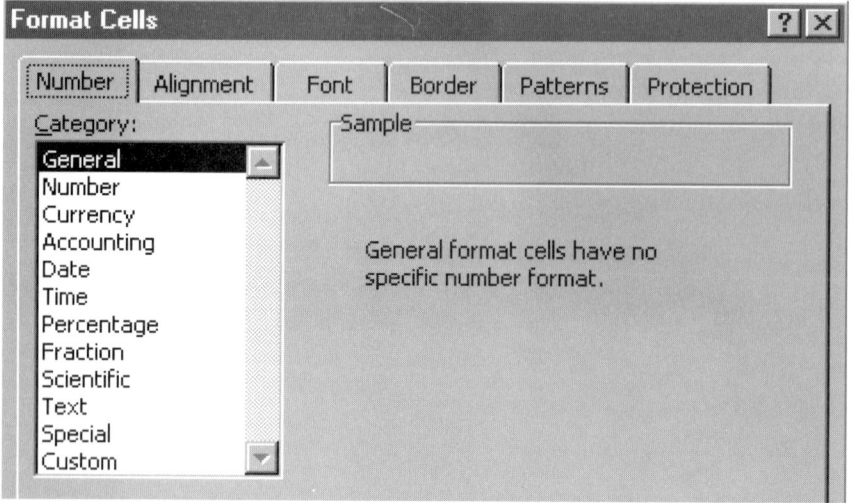

Using the Number tab you can choose how you wish your data to be displayed. 1/3 was entered into each of the cells A1 to A9 below:

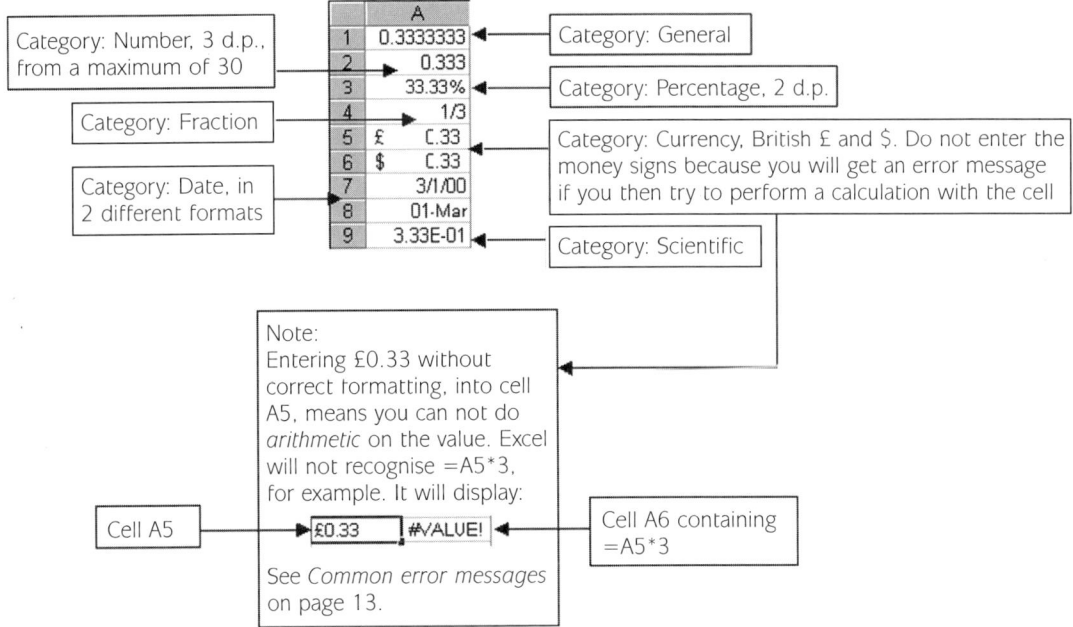

Note: If you try and perform an operation on data that has been incorrectly formatted, Excel will return the error #VALUE. To enter £0.33, you must enter 0.33 and then format the cell for currency.

The *Font*, *Border* and *Patterns* tabs allow you to add colours, shading and grids to your spreadsheets.

You can use the alignment tab to help format text in your headings. In the diagram below, 'Term in Series' was entered into cell A1 and 'Fibonacci Sequence' was entered into cell B1. Without widening the columns there is an overlap:

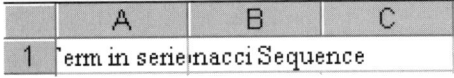

Widening the columns and rows is not always helpful as it means you see less on screen.

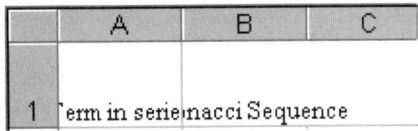

To wrap the text within the cell, highlight cells A1 and B1, click on Format and Cells, and then select the Alignment tab and make sure the Wrap text box is checked:

When you click OK your spreadsheet will be formatted correctly:

	A	B
1	Term in series	Fibonacci Sequence

Protecting your worksheet

The Excel files on the CD-ROM are protected so that students cannot alter any data, charts or text on the sheet. To protect a completed sheet, click on Tools then Protection then Protect sheet... Enter a password when you are prompted and all the cells will be locked. If you wish to leave cells 'open' to enter data, then, before protecting the worksheet, highlight any cells that you want to be left 'open', click on Format and Cells, select the protection tab, and remove the tick on Locked:

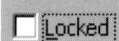

To unprotect the whole sheet, click on Tools then Protection then Unprotect sheet... Re-enter the password that you used to lock the sheet and it will be open for editing again.

Using the Excel chart wizard

You can use the Excel chart wizard to draw graphs based on data from a spreadsheet. The following data shows cars sold and part exchanged at a garage during one year.

	A	B	C
	Month	Cars sold	Cars part exchanged
1			
2	Jan	37	12
3	Feb	42	15
4	Mch	41	14
5	Apr	55	27
6	May	60	12
7	Jun	59	17
8	Jly	68	9
9	Aug	65	26
10	Sep	72	29
11	Oct	68	29
12	Nov	59	24
13	Dec	71	25

You can draw a bar chart of sales per month by highlighting the data in columns A and B and clicking the chart wizard button to take you step-by-step through the process of creating the graph.

To highlight cells next to each other you can simply click and drag the mouse to create a rectangle that covers all the cells you need. To highlight cells that are not next to each other, highlight one group of cells and then hold down the Ctrl key while you click and drag to highlight the others.

The chart wizard button:

If you highlight the column titles as well, the chart wizard will add the headings for data columns to the legend on the right hand side of the chart.

Common error messages

You will probably see a number of error messages as you use Excel. The three most common that you will come across are:

#DIV/0   A student has divided by zero or an empty cell.

########## The cell is not wide enough to display the number

#VALUE   A cell containing text has been used in a formula (eg = B1+A2 where A2 contains a number and B1 contains text.

Printing from Excel

A potential problem in printing from Excel is the amount of data in a worksheet. Just clicking the print button can waste a great deal of paper. Always ask students to do a print preview before *any* printing. Two methods for printing are suggested below.

To print a selection of a sheet, highlight the cells to print and click File followed by Print. Make sure that the Selection radio button is checked and then click the OK button to print.

Alternatively, click File from the menu bar followed by Page Setup... and select:

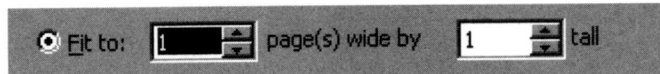

If there is a lot of data on the sheet the second method can produce very small text!

A guide to using Omnigraph

The graphing package Omnigraph is very powerful and easy to use. The activity sheets in the Pack contain hints in the margin box where appropriate. Omnigraph is also helpful when used through a data projector to demonstrate topics in all Key Stages of Mathematics. Some ideas for this type of use are given on pages 44–46.

The Omnigraph window is shown below and holding your mouse pointer over any of the buttons will display their use.

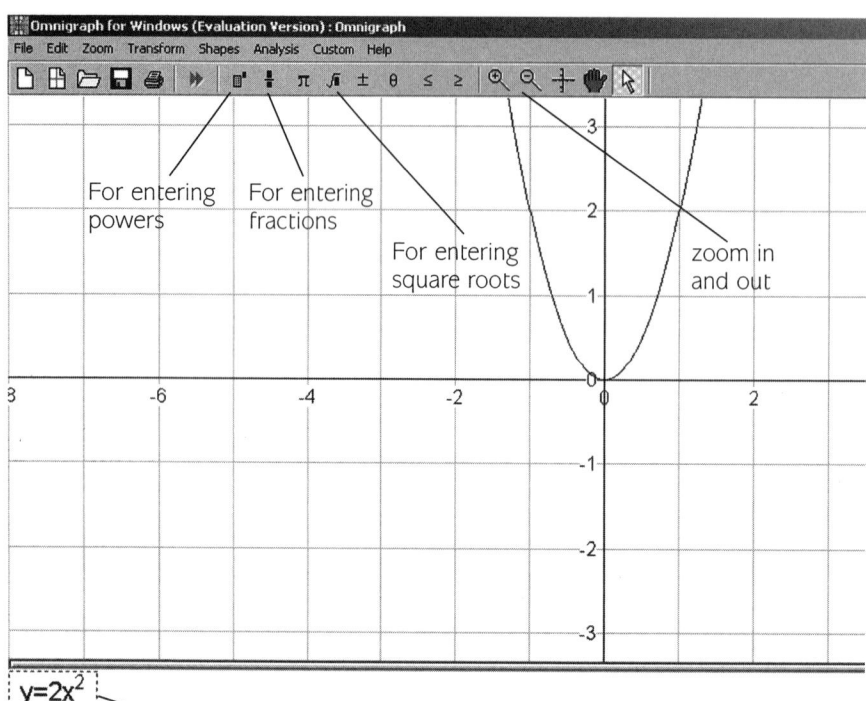

For entering powers

For entering fractions

For entering square roots

zoom in and out

$y=2x^2$

- Enter your equations here.
- You do not need gaps in the equation.
- Press the Enter key to produce the curve.

Producing trigonometric curves

- When using degrees

On the menu bar click <u>Z</u>oom and then <u>T</u>rig scales and then enter your curve:

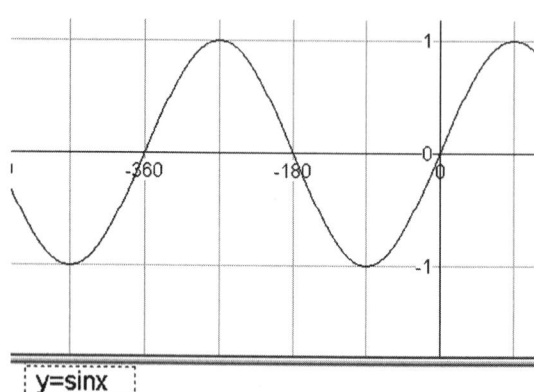

y=sinx

- When using radians

You must convert to radian measure *before* selecting the Trig scales. To convert to radian measure click <u>C</u>ustom and then select the Radians radio button and click the OK button then convert to Trig scales.

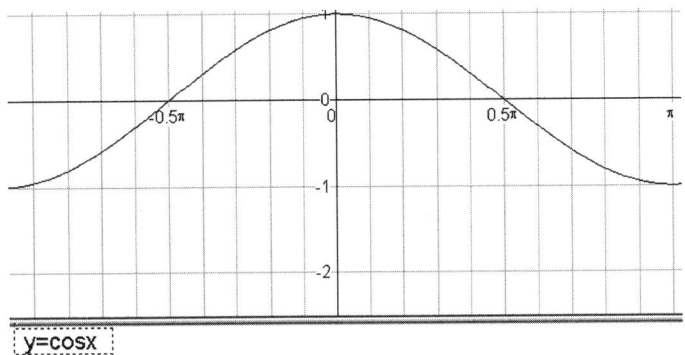

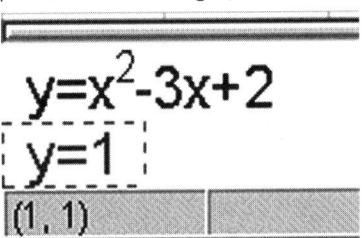

The coordinates of the mouse pointer are displayed below the equation box when the mouse pointer is over the graph window.

Click <u>F</u>ile and <u>C</u>lear curves to remove any trigonometric curves and then <u>Z</u>oom and <u>R</u>eset scales before entering Cartesian equations.

Using the gradient function

The gradient function provides a good introduction to the 'A'-level topic of Differentiation. Remember to change to radian measure if you want to view the associated gradient functions of the trigonometric functions.

To produce the gradient function for cos*x*, click on <u>A</u>nalysis and then <u>G</u>radient. The gradient function, – sin*x*, will then appear on screen:

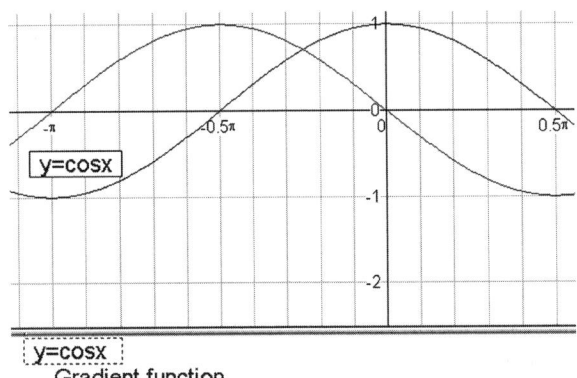

Gradient function

Using the dynamic constant editor

This is very useful when, for example, demonstrating function transformations. Begin by defining any constants used in your function. Enter $a = 0$ and then $y = (x - a)^2$ to produce the graph of $y = x^2$:

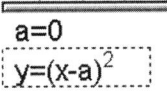

Now bring up the dynamic editor box by clicking <u>E</u>dit and then Dynamic constants or by clicking the button which is added to the menu bar after the constant has been defined. Change the step to 1:

Click this button to step through different values of *a* and watch the curve translate on screen.

You must close down the Dynamic constant editor box before you can use the main menu again.

Printing from Omnigraph

It is a good idea if students add their name to any curves they wish to print out. To do this click on <u>C</u>ustom and then Comment box:

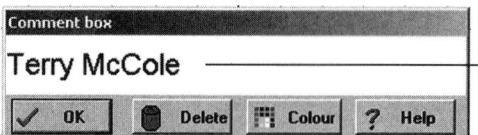

Students can enter their name here.

Clicking the OK button produces the text on screen which can then be moved to an appropriate position:

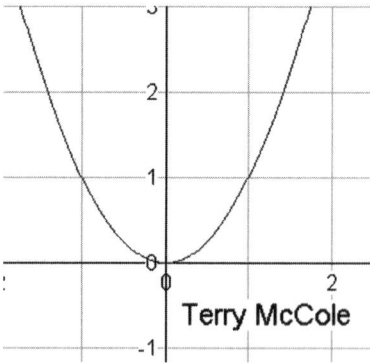

Other Features

The handbook with Omnigraph provides an easy to use guide for all the other facilities available. These include the plotting of polar and parametric curves, the plotting of coordinates and the transformations of shapes. See, for example, Activity SSM9 on page 133.

A guide to using MSWLogo

MSWLogo is one of the many versions of logo software that enables you to programme in a language and control a 'turtle' on screen.

The main screen is split up into three sections:

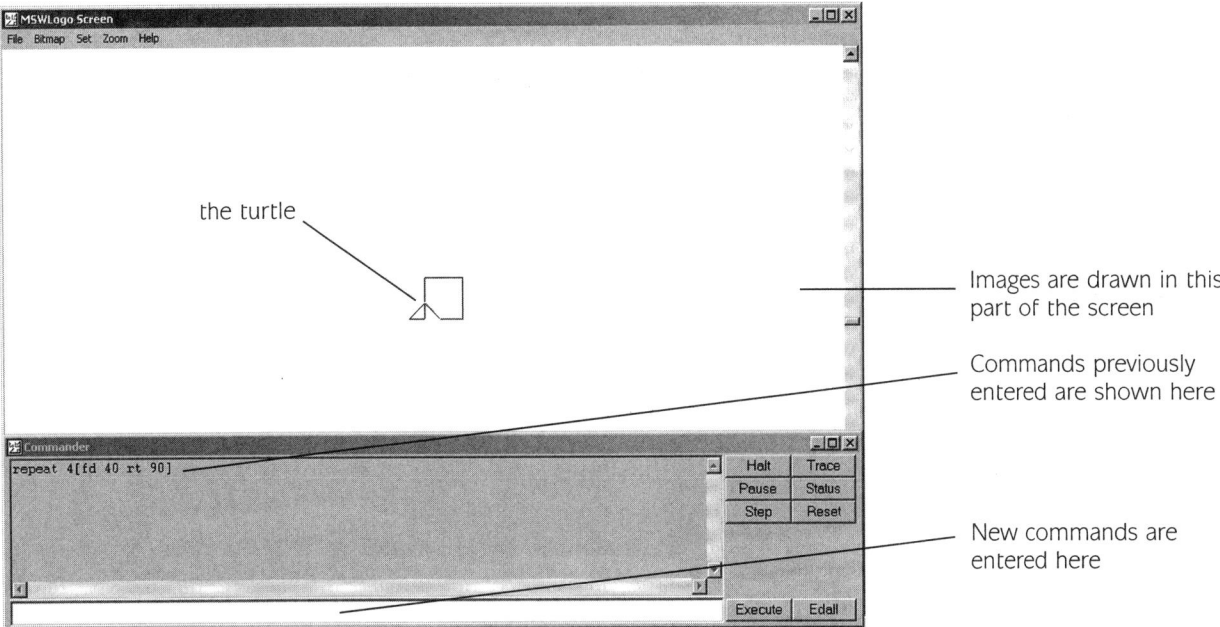

the turtle

Images are drawn in this part of the screen

Commands previously entered are shown here

New commands are entered here

Some common commands

To enter a command you type the necessary code, press enter, and the turtle will follow the instructions given.

Here are some common commands:

forward n – Moves the turtle forward *n* units. Enter `forward 60` or `fd 60` to move the turtle 60 units forward
back n – Moves the turtle backwards *n* units. Enter `back 45` or `bk 45` to move the turtle 45 units backward
left n – Turns the turtle *n* degrees to the left. Enter `left 90` or `lt 90` to turn the turtle 90° left.
right n – Turns the turtle *n* degrees to the right. Enter `right 60` or `rt 60` to turn the turtle 60° right.

Here are the commands to draw a square of side 50 units:

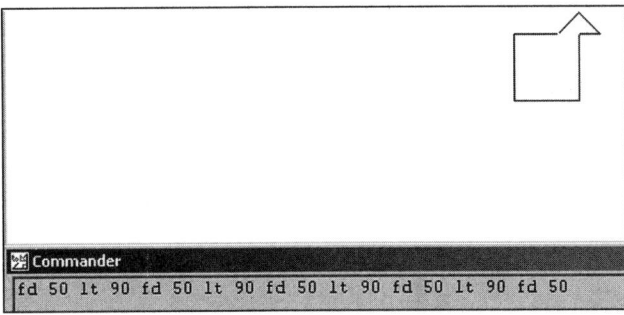

Commander
`fd 50 lt 90 fd 50 lt 90 fd 50 lt 90 fd 50 lt 90 fd 50`

See page 19 for a better way to create a square.

Some other Useful Commands are:

clearscreen or **cs** – clears the screen and returns the turtle to the origin.
penup or **pu** – lifts the pen off the paper and prevents it drawing a line.
pendown or **pd** – puts the pen down again.

The Menu bar

Bitmap

With the bitmap menu you can load and save bitmaps, or pictures, on your screen. You can also print out your current screen display.

Set

Some of the logo commands are more easily obtained from using the menu bar. Click on <u>S</u>et and then <u>P</u>enColor to change the pen colour:

Zoom

This menu allows you to zoom in and out of your picture. This may be useful for the Imagine worksheet, activity SSM8 on page 131.

Procedures

Entering the instructions for the square above creates problems if you make a mistake as you will have to type in all the commands again. It is better to create a procedure. A procedure is a list of commands which you create in the **Editor**. To create a procedure use the **edit** command.

Enter `edit "square` in the commander:

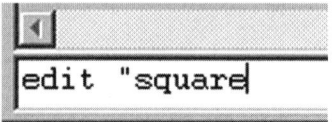

This produces the Editor window. Enter the code shown below.

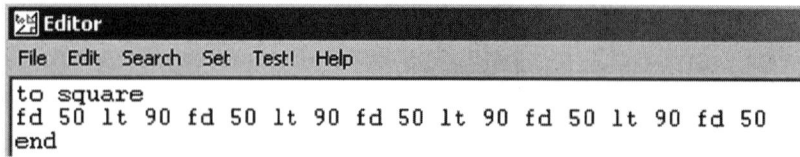

The name of the procedure is preceded by to. Finish the procedure with end.

After entering the code, click on <u>F</u>ile then S<u>a</u>ve and Exit.

Now type `square` in the commander box and press the enter key to draw a square:

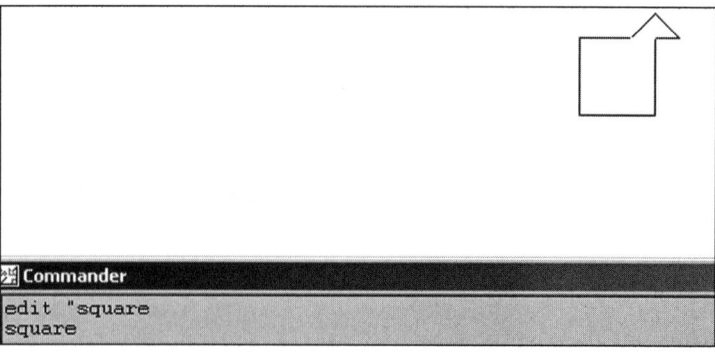

Any procedure can be created in this way. To edit a procedure you can type `edit "square` in the commander box again or select File and Edit from the menu bar.

Economy of Language

The procedure to produce the square repeats the same command 4 times. To avoid this you can use the *Repeat* command.

The square procedure can be replaced by:

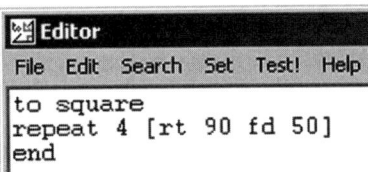

```
Editor
File  Edit  Search  Set  Test!  Help
to square
repeat 4 [rt 90 fd 50]
end
```

The repeat command has two parts:
■ the number of times to repeat, 4 for the square
■ the commands to repeat, contained in square brackets.

Calculations

Logo will perform calculations for you. To do this you use *variables*.

To give a value to a variable you use the quote marks and the command `make`. In this example we are giving a variable, *length*, the value of 50.

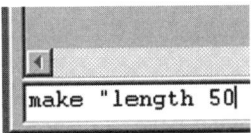

```
make "length 50
```

To check your value you can use the show command. Use a colon preceding the variable to do this:

```
show :length
```

To use variables in procedures you specify them after the procedure name. This example creates a generic procedure to draw any regular polygon.

Enter the following procedure:

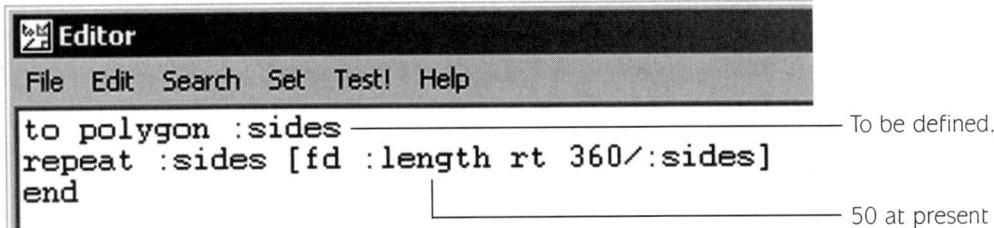

```
Editor
File  Edit  Search  Set  Test!  Help
to polygon :sides ──────────────────── To be defined.
repeat :sides [fd :length rt 360/:sides]
end            └─────────────────── 50 at present
```

Enter `polygon 5` in the commander box to produce a regular pentagon of side 50:

Enter `polygon 10` in the commander box to produce a regular decagon of side 50:

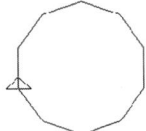

Enter make "length 30 polygon 6 in the commander
box to produce a regular hexagon of side 30:

Before you exit from MSWLogo you should save any procedures you
have written. All the procedures written in any one session will be
saved in the same file. Click on File and then Load to access the
procedures again.

Moving the position of the turtle

When the program starts the position of the turtle is its origin [0 0].
Note the square brackets and no comma. To move the turtle to
[−40 80] type:

```
setpos [-40 80]
```

The code:

```
pu setpos [-40 80] pd
```

will move the turtle without it drawing a line and put the pen down
ready for any further instructions.

Printing

It is a good idea for students to add their name before they print.
The code label "Debbie will produce the text Debbie at the
position of the turtle [use the command setpos to move the turtle to a
suitable position]:

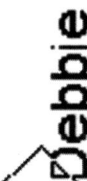

Demonstrations

MSWLogo has a number of demonstrations which show the full
potential of the software and also a user friendly Help system.

Errors

It is not possible to delete any incorrect lines on screen. However, if
you Set the Pencolor to white and reverse the last instruction, you
will delete the last line drawn. Do not to forget to change the pen
colour back from white again!

A guide to using The Geometer's Sketchpad

The dynamic geometry package The Geometer's Sketchpad is a lovely way to discover the properties of Euclidean geometry. It is fun to use and provides powerful images to enhance the learning and understanding of geometry. The pupil activity sheets in the Pack contain hints in the margin box where appropriate.

Getting started...

When the program opens you will see this screen:

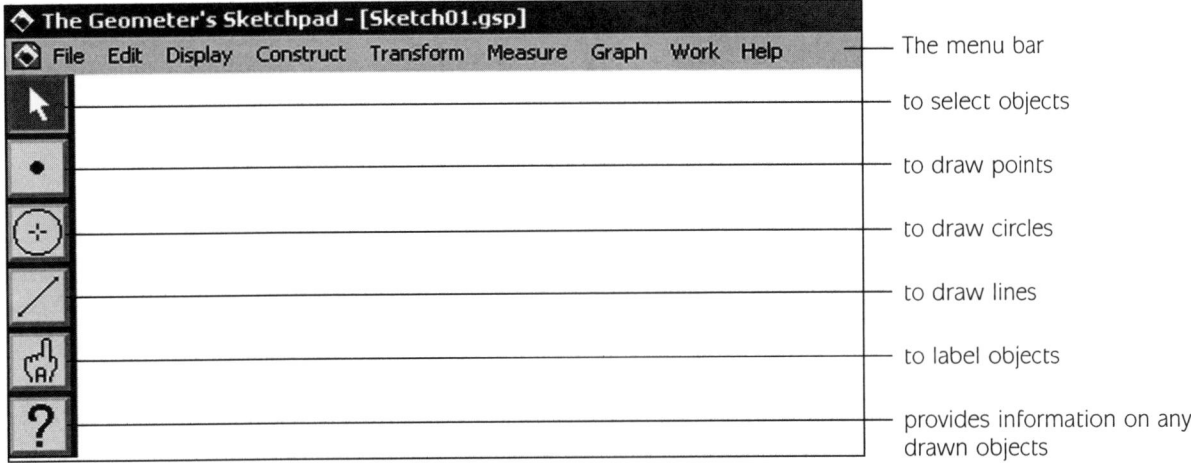

To let the program automatically label objects as they are drawn, click on <u>D</u>isplay and <u>P</u>references on the menu bar and tick the Autoshow Labels for <u>P</u>oints, <u>S</u>traight Objects and C<u>i</u>rcles.

In the same Object Preferences dialogue box, change, if necessary, to the following Units and Precision:

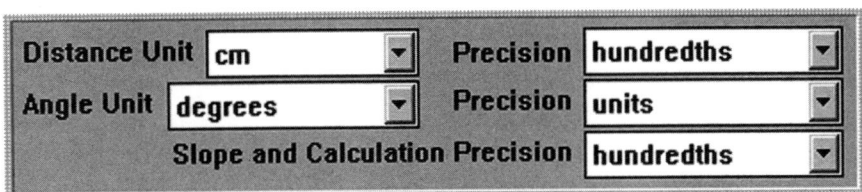

Drawing and measuring

To draw a line:

■ Click on the line tool button on the drawing toolbar and drag out a line in the sketchpad window. The line is automatically labelled and selected:

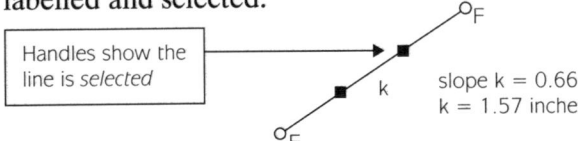

In Geometer's Sketchpad, the gradient of a line is called the slope.

- If you wish to select another object, click the top button on the drawing toolbar and then click on the required object.

Don't forget that you need to have the arrow in the menu bar selected.

To measure the gradient of a line:

- Select it and click Measure, Slope.

To measure the length of a line:

- Select it and click Measure, Length.

To select more than one object or point:

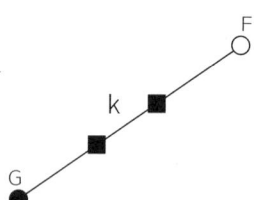

- Hold down the Shift key on your keyboard and click on each object in turn.

- To measure the size of the angle EFH the points E, F and H need to be selected in that order before clicking Measure and then Angle. Deselect the points E, F and H first if necessary.

The point G and the line k are both selected but not the point F.

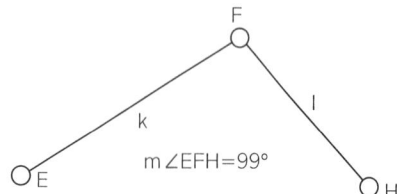

Don't forget that you need to have the arrow in the menu bar selected.

If you drag the point H to change the angle, the program automatically updates the size of ∠EFH.

- To produce the calculator, click Measure and then Calculate on the menu bar.

To use the calculator to sum two angles, click the *first* angle, m∠CDB = 45° to make it appear in the calculator, followed by the *add button* on the calculator, then click the *second* angle, and finally the *OK button* on the calculator

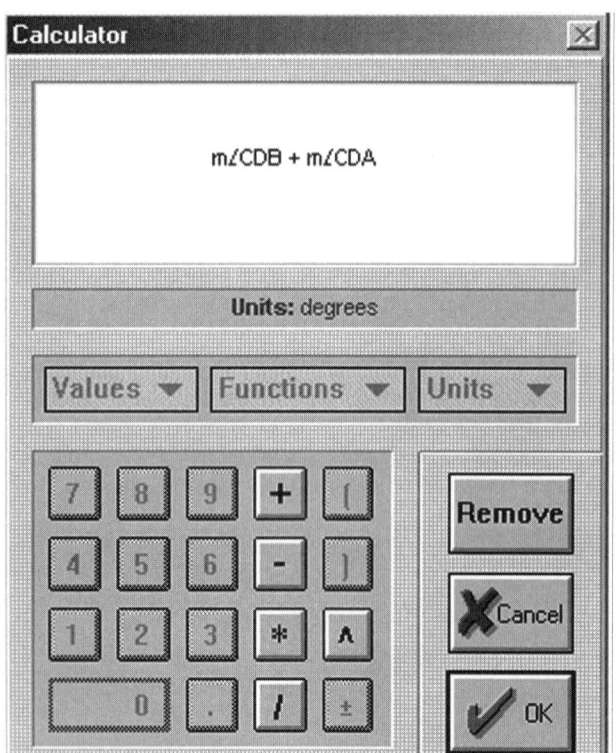

The angle sum is displayed on the screen.

m∠CDB = 58°
m∠CDA = 122°
m∠CDB + m∠CDA = 180°

- To show the co-ordinate axes click Graph, Create Axes.

Samples

The software comes with samples of what can be achieved. The file sketch/samples/presentn/sinwaver.gsp, for example, produces an animation of the sine curve showing its relationship to a circle. It would provide a splendid introduction to the sine curve if demonstrated via a data projector.

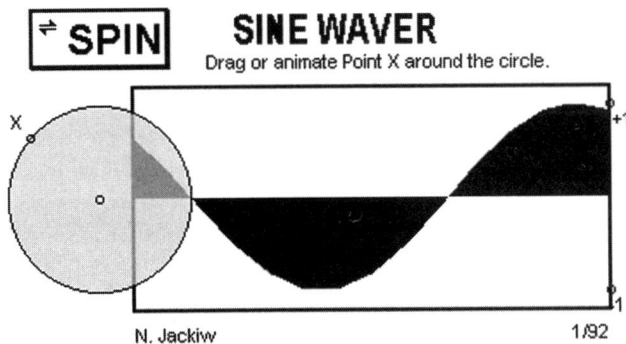

There is also a puzzle based on Pythagoras' Theorem which provides an excellent interactive student activity:

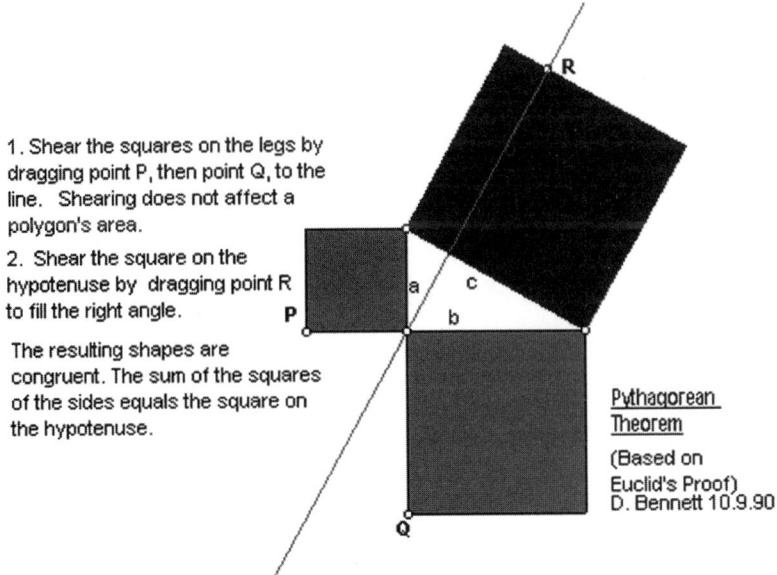

It is one of four activities based on Pythagoras' Theorem and one of many other activities ranging in diversity from the two above, to an investigation on how a, b, and c affect the shape of the curve $y = ax^2 + bx + c$, to an investigation into the flight of a projectile, to a demonstration of the familiar 'Max Box' problem....

Note:
- when using the activity sheets the labels you get on screen may not match the ones on the activity sheet.

As the software says:

"Enjoy!"

Searching the internet

Using the internet or the World Wide Web (WWW) can be both the most rewarding and most frustrating experience! There are some wonderful sites on the internet that can generate interesting and informative mathematics lessons, but finding them can be a problem as there are also many, many poor sites!

Below are a number of ideas that you can use in order to get the most from such a vast wealth of information.

In lessons

Taking a class into the ICT room and asking them to search the internet can be a waste of lesson time. Pupils will inevitably stray from your chosen subject and they may not extract the necessary information on any appropriate sites located.

You could place links to your favourite Mathematics pages on your intranet or school website.

It is much more productive to find the sites before the lesson and to ask pupils to stick with these sites. If you are familiar with the content of any chosen site then you will know exactly what your pupils are aiming for.

Searching the WWW

*The content below is based on the search engine Yahoo (*www.yahoo.com*).*

If you know exactly what you're looking for or just have a general idea, try using a search engine. Specify a keyword or set of keywords, and the search engine will check its entire database to find websites that match the keywords you provide.

The Yahoo query box

The first page returned to you will be a list of matching Yahoo categories followed by a list of suggested sites.

- **Use double quotes around words that are part of a phrase.**
  For example: "great barrier reef"
  This means the engine searches for the full text rather than *great* or *barrier* or *reef*.

- **Specify words that must appear in the results.**
  Type + in front of words that *must* appear in result documents.
  For example: sting + police

- **Specify words that should not appear in the results.**
  Type – in front of words that *must not* appear in result documents.
  For example: python – monty

Other search engines:
- www.uk.altavista.com
- www.lycos.co.uk
- hotbot.lycos.com
- www.askjeeves.com
- www.google.com

A search engine will begin by finding all the keyword matches and then sort the results.

Exact matching
- **Sting**chronicity – Sting & Police Lyrics

Close matching
- **Sting**chronicity –
- FAQ – The **Police**

Related matching
- **Sting**mania – Lyrics

Unrelated matching
- Compliance Checks are a Miserable Failure – details why **sting** accomplishes nothing

  Law Enforcement

These two sites, which have nothing to do with music, appear because the words 'police' and 'sting' appeared in the query box.

Sometimes you can use the categories already compiled in the search engine. Here the Yahoo 'Entertainment' was used to find sites relating to 'Sting'.

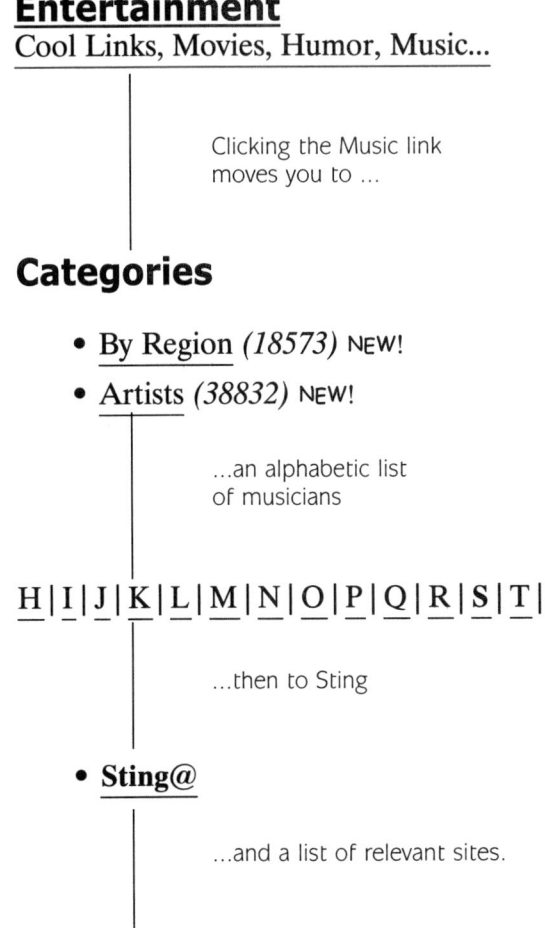

Entertainment
Cool Links, Movies, Humor, Music...

Clicking the Music link moves you to ...

Categories

- By Region *(18573)* NEW!
- Artists *(38832)* NEW!

...an alphabetic list of musicians

H | I | J | K | L | M | N | O | P | Q | R | S | T |

...then to Sting

- **Sting@**

...and a list of relevant sites.

- Berklee College of Music: Sting – text of his 1994 graduation speech
- Brand New Day Tour – official site for Sting's 1999/2000 world tour
- Calling Sting – link list
- **Police@**

Found a good site?

When you have found a good internet site you should keep a record of its address. In Internet Explorer you can add it to your favorites, by clicking on Favorites and then Add to Favorites:

Note the American spelling of favourites!

You can then add the favourite to an existing Folder or make a New Folder...

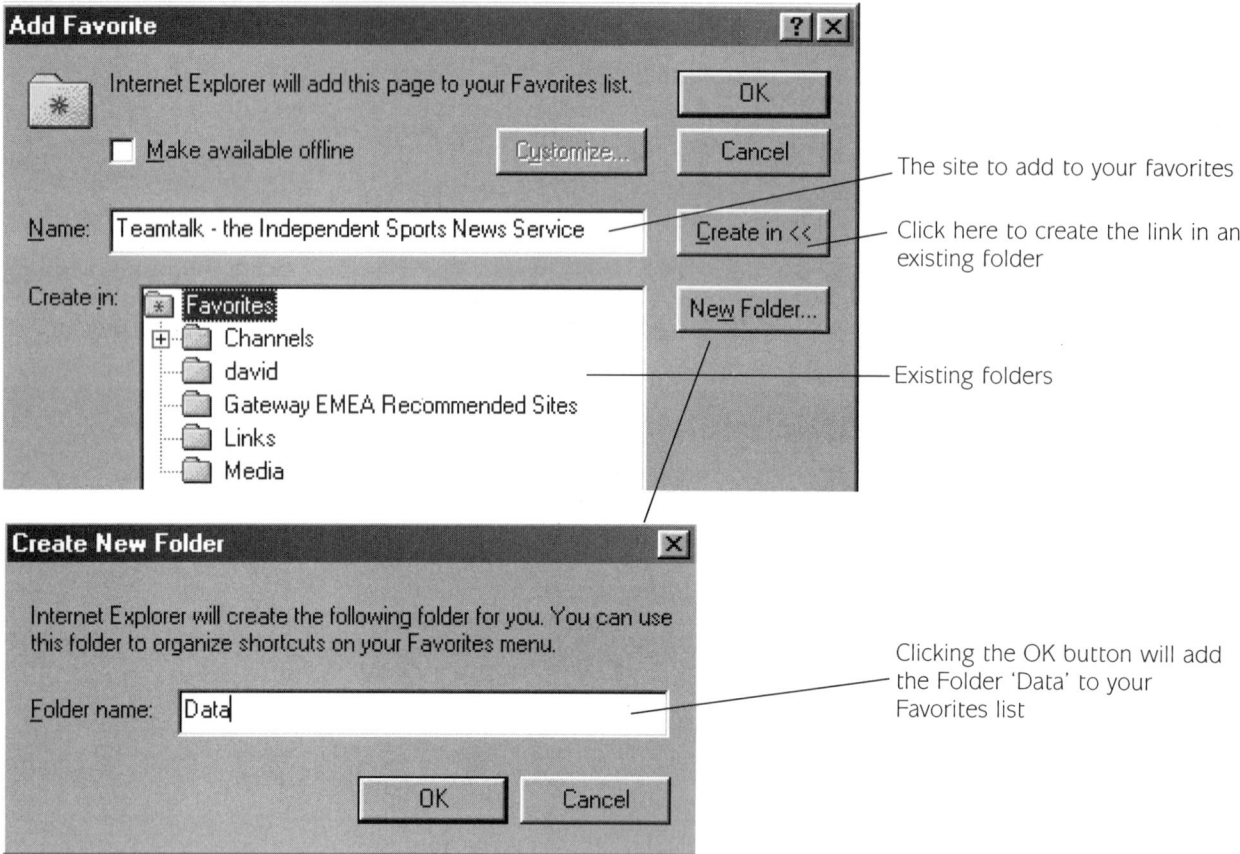

The site to add to your favorites

Click here to create the link in an existing folder

Existing folders

Clicking the OK button will add the Folder 'Data' to your Favorites list

Mathematics

An excellent start for any mathematical search is:

http://mathworld.wolfram.com

which contains many links to anything mathematical. Type pi in the search box and off you go...

Printing from the internet

Hitting the print button on an internet site can waste a lot of paper. If you or your students want to obtain some information from a site it is best to highlight and then copy the required text and paste it into a word processing document. Right-clicking on an internet image allows the image to be copied before pasting it into the document.

Pythagoras of Samos is often described as the first pure mathematician. He is an extremely important figure in the development of mathematics yet we know relatively little about his mathematical achievements.

Website information

The internet also provides homework and revision sites for students. Some useful sites are given below:

Amazing number facts
(http://users.zetnet.co.uk/madras/maths/amazingnofacts/index.html)

BBC Schools Online (http://www.bbc.co.uk/education/schools)
This curriculum-based site offers a variety of interactive learning resources for children at home and at school. Some activities are linked to BBC programmes for schools.

Becta (http://www.becta.org.uk/index.cfm)
Information for teachers.

Birmingham Mathematics
(http://atschool.eduweb.co.uk/ufa10/index.html)
Good links, puzzle of the week, lesson starters etc.

Count on (http://www.mathsyear.com)
Puzzles, information, tutorials, tests etc.

Doctor Math (http://mathforum.com/dr.math/)
Even though this is a US site, it is still useful as pupils can post queries or homework questions to Dr Math and receive answers. There is also a library of previous questions that can be searched.

Eric Weisstein's World of Mathematics
(http://mathworld.wolfram.com)
An excellent site to start any mathematical search.

ERIC WEISSTEIN'S
world of
MATHEMATICS

The site contains more than
- 10,000 entries
- 87,000 cross references
- 4,000 figures
- 100 animated graphics
- 950 Java applets.

It also has a latest news link.

NOVEMBER 24, 2001

**MATHWORLD™
HEADLINE NEWS**

New Mersenne Prime (Probably) Discovered
A new Mersenne prime, the largest discovered to date, appears to have been found.

Exeter University Centre for Innovation in Mathematics

(http://www.ex.ac.uk/cimt/)
Teaching site containing resources, links, puzzles a dictionary of units with conversion calculator, a data bank and much more.

Logo art gallery (http://www.geocities.com/CollegePark/Lab/2276/)

Lots of pictures created in Logo. Encourage pupils to create a school Logo art gallery and put it on the school website.

Homework High

(http://www.homeworkhigh.co.uk/maths/frame.html)
Pupils can post queries or homework questions here and receive answers from one of the sites experts. There is also a library of previous questions that can be searched by keyword or browsed by age range.

NGfL website (http://www.ngfl.gov.uk)

The NGfL site is a good starting point for professional development and research. Both students and teachers can make use of the NGfL website's content.

NRich (http://www.nrich.maths.org.uk)

An online mathematics club with puzzles and Logoland.

Number facts (http://www.nottingham.ac.uk/education/number/)

About the numbers in today's date, good for lesson starters.

Online encyclopedia of integer sequences

(http://www.research.att.com/~njas/sequences/)
Everything you would wish to know about sequences.

Yahoo (http://dir.yahoo.com/science/mathematics/education)

A good start for any mathematical search.

Using an intranet

An intranet uses the same technology as the internet but is usually only available within school. However it is possible to allow use of the intranet off site, allowing students and staff access to its contents both at home and at school.

Ideas for your intranet

The school intranet can provide information, policies, worksheets, internet links, handbooks, advice, etc. to both students and staff. Everything can then be accessed from a workstation at any time during the school day.

If your school does not have an intranet you can purchase them 'ready made' and then customise them to suit your own school needs.

Below are some ideas of what can be put onto your intranet.

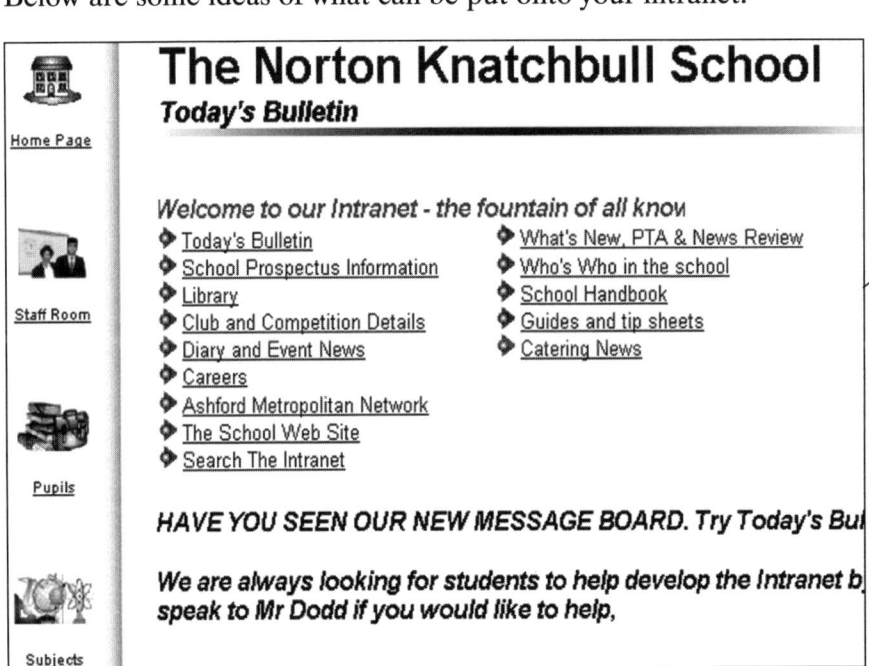

An example of a school intranet page containing information about the school, students and staff.

The Mathematics Menu

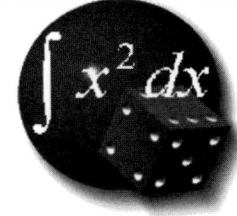

Ideas for the mathematics menu.

Internet Resources Menu

This page contains hyperlinks to websites useful for both students and staff. The links are best accessed from a worksheet on the intranet where specific tasks are included for the student.

General Interest
- MathsWorld
- Escher World
- The Origins of Algebra
- History of Mathematics
- Visual Dictionary of Plane Curves
- Pi
- History of Mathematicians
- University of Essex Centre for Innovative Mather
- Birmingham Resources Page
- More Escher

Worksheets

You can create the worksheets using a word processing package and then transfer them to the intranet.

Year 7
- Year 7 Generating Series 1
- Year 7 Generating Series 2
- Year 7 Data 1
- Year 7 Data 2
- Year 7 Data 3

Below is part of a worksheet providing an online use of the 24-hour clock and the arithmetic of time.

Formatting the background and adding pictures make the sheet more interesting for the students.

Press Here to Link to:

British Rail Timetables

Part of a worksheet with a direct link to online railway timetables.

It is important to provide any internet links as too much lesson time can be wasted if students are allowed to surf for the necessary information themselves.

If each member of the department created one worksheet a week you would soon build up a large resource.

Ideas for creating an interactive worksheet containing work on averages and scatter graphs

n Find a suitable internet site in advance:
www.teamtalk.com

■ Create a worksheet with links to the necessary site(s).

■ Work through the questions to check that everything is OK.

■ Give details of exactly what has to be done:

1 Go to the site of each Premier League team and use the PLAYERS link to find the following statistics about their goalkeepers: age, weight, height.

2 Launch Excel and use it to find the mean average age of all the goalkeepers.

3 Create a scatter graph of weight against height and comment on the type of correlation.

■ Provide any necessary hints.

The formula
= average (A2:A32)
calculates the mean of all the data in cells A2 down to A32

■ Provide the National Curriculum Levels.

Level 5 (Mean) H5d

Schemes of Work

You could place your schemes of work on the intranet so that students are clear about what they are expected to know.

Create a *secure* staff area if there are pages you wish to 'hide' from your students, for example, your department handbook.

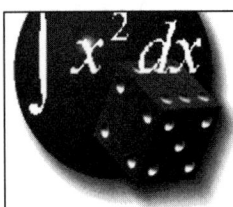

GCSE Higher Schemes of Work

In addition to the work below the following terms include:
Term 1: practice coursework investigation (fortnight of mocks)
Term 3 Exams
Term 4 coursework investigation (fortnight before half term)
mock exams
Term 5 coursework investigation (fortnight before half term)

Timing	Topic	Reference	Notes
Year 10	Arc length and sector area	19.1 pg 334 Y10arcs2.doc y10arcs.doc	sets 1 and 2 discuss radia their us NKS softw
	Volume and surface area	19.2 pg 342 10volume.doc	NKS softw
	%ages (Percentage change required for	5.1 - 5.5 pg 91 10perwks.doc y10perc.doc	include inve age calcula

This work scheme lists the documents accessible on the intranet.

National Curriculum Attainment Levels and Programmes of Study

These are useful for both students and staff. It may even encourage your department to read them and print out copies when OFSTED are arriving!

Homeworks

You can include a bank of word-processed homeworks. This will be particularly useful if the intranet can be accessed at home; or they can be posted on the school internet site. Copies may be edited at any time and whole class or individual copies may be printed.

Turn your ideas into intranet pages.

Examination Papers

Turn past examination papers into intranet pages. These are especially useful if they also contain the examiner's comments, hints, tips and solutions. If your intranet is accessible from home then this can be a valuable revision resource for students on study leave.

The school can purchase CD-ROMs containing examination questions. With a whole school licence a lot of time and effort can be saved by transferring the questions to the intranet.

Coursework Guidance

Include any external coursework guidelines and include extracts of projects showing exemplary work from former students.

Creating high-quality worksheets

It is possible to create professional looking worksheets using a word processing package. The worksheets can be amended easily and saved for future use. The use of Microsoft Word 97 is demonstrated here.

Most of the ideas demonstrated here can be done in Word versions 2 and 6. See pages 40–41 for some differences.

Inserting mathematical symbols

To insert a mathematical symbol click Insert and then Symbol on the main menu to display the Symbol dialogue box:

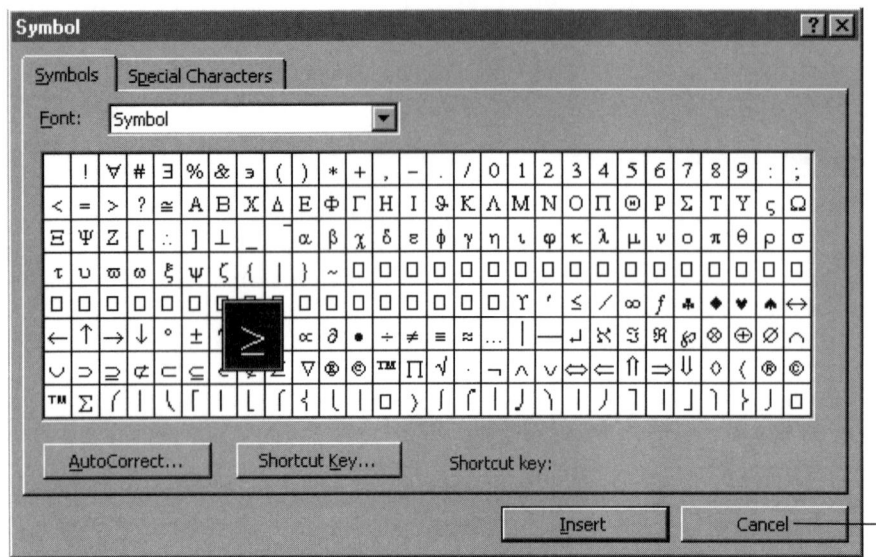

The Close button replaces the Cancel button after Insert is clicked.

To choose a 'greater than or equal to' sign, click the symbol, then click Insert and Close. The ≥ will then appear in your document.

Building more complex expressions

When more advanced mathematical expressions are required, the Microsoft Equation Editor can be used to create more complex expressions. It requires no more than 30 minutes to become proficient and you will be able to produce expressions like these:

$$\int_0^5 x^2 dx = \left[\frac{x^3}{3}\right]_0^5 = \left(\frac{125}{3}\right) - (0) = 41\frac{2}{3}$$

To open the Equation Editor, click Insert then Object. The Object box will then appear:

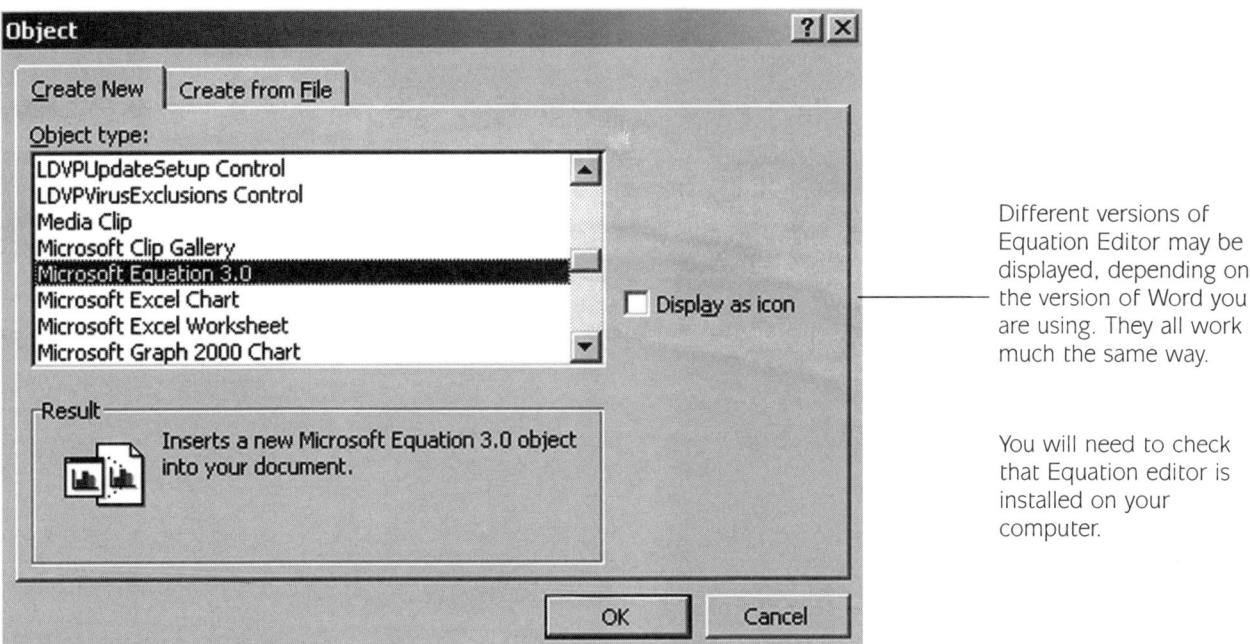

Different versions of Equation Editor may be displayed, depending on the version of Word you are using. They all work much the same way.

You will need to check that Equation editor is installed on your computer.

Double click on Microsoft Equation. If it does not appear on your list, ask your IT coordinator to install it for you.

This produces a dialogue box similar to the one below.

Each of these buttons produces a drop down menu containing a variety of options.

with an associated toolbar.

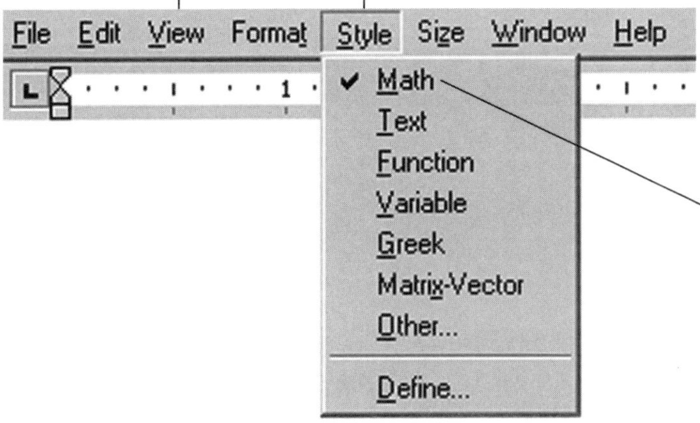

Style is the most useful option. Choose Text for normal typing and Math for mathematical typing. Switch between the two as necessary.

With Text selected you can type as normal.

With Math selected, all text, apart from mathematical functions like sin and ln, are in Italic.

It may be necessary to change the font to be the same as the one used in your Worksheet. This may be done by clicking Style and Define and changing all except L.C. Greek, U.C. Greek and Symbol.

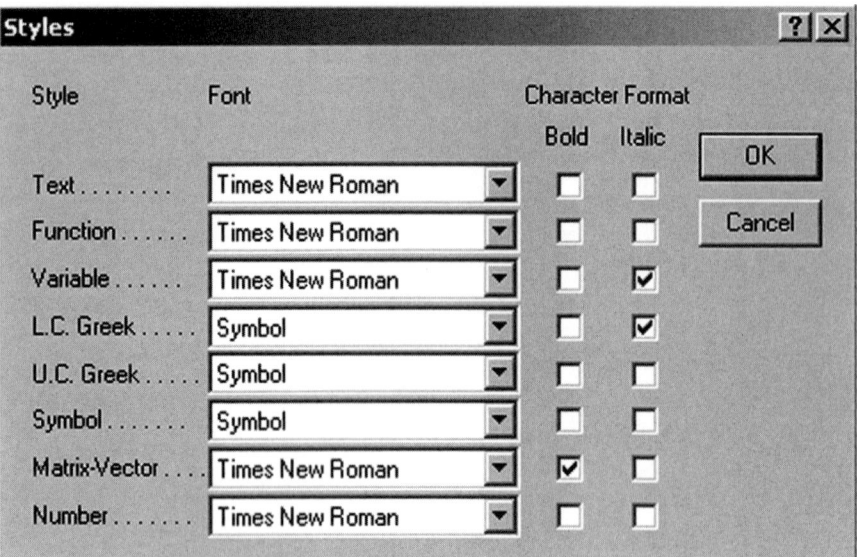

A window will appear in your word document ready to accept the Equation Editor typing.

Try the practical session on Equation Editor on pages 41–43.

It is advisable not to put more than one page into Equation Editor, as page breaks can prove problematic.

Adding diagrams to your worksheet

You can add diagrams into your worksheets in a variety of ways. Three methods are suggested here.

Using the Microsoft Word Drawing Toolbar

Click View, Toolbars and then select Drawing to add a drawing toolbar similar to

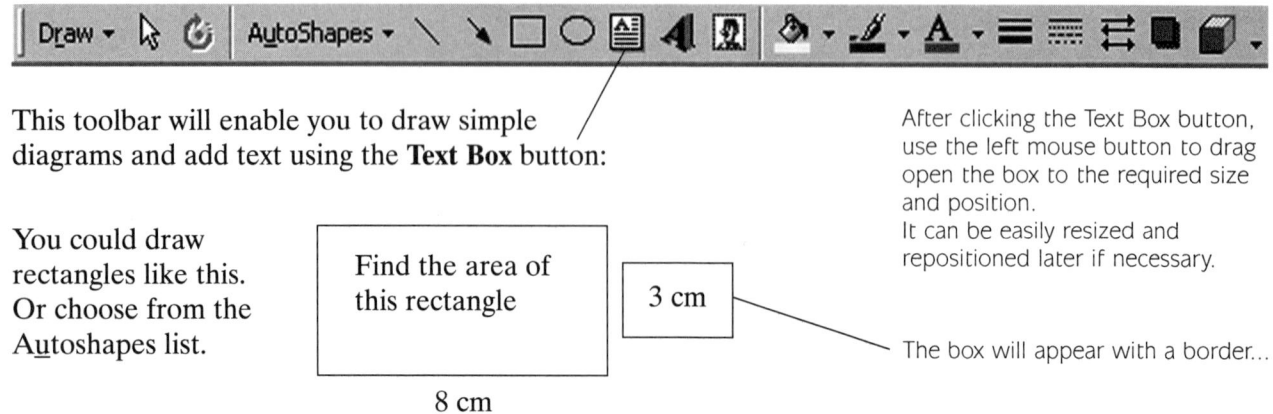

This toolbar will enable you to draw simple diagrams and add text using the **Text Box** button:

After clicking the Text Box button, use the left mouse button to drag open the box to the required size and position.
It can be easily resized and repositioned later if necessary.

You could draw rectangles like this. Or choose from the Autoshapes list.

Find the area of this rectangle

3 cm

The box will appear with a border...

8 cm

To remove the border, double click on the text box border, click on Line, Color: and select No Line. Also select: Fill Color and No Fill as this will help the positioning of the box.

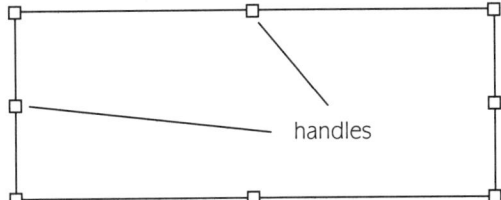

handles

To add text inside a shape, see page 40.

Using a specific drawing package

When more complicated drawings are required, it is best to use a specialist drawing package such as Micrografx or Windows Draw 3.0.

In Windows 3.x, switch between the two packages as required by holding down the Alt key on your keyboard while you press the Tab key.

It is possible to have your Drawing and Word Processing packages open at the same time.

A depressed button indicates the program window currently visible on your screen.

In Windows 95 and any higher version, use the bar at the base of your screen to switch between packages.

With a little practice, you can produce diagrams like the one below, and you can save them and build a bank of them for further use.

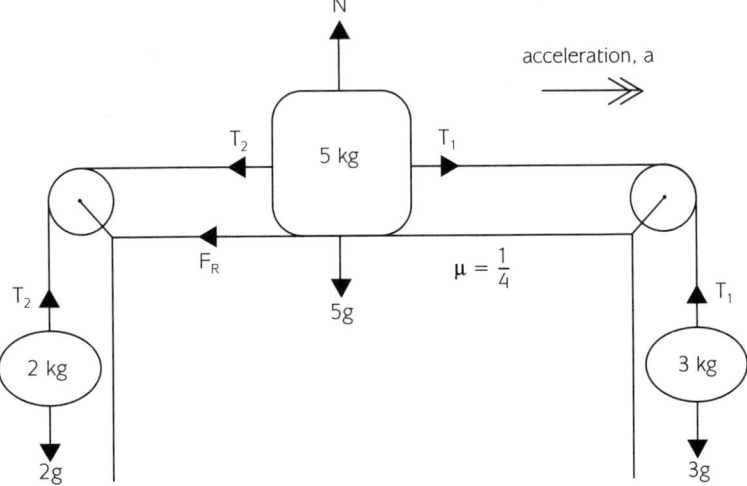

Using a mathematics drawing package

FX Draw is an example of a package containing many mathematical images, which you can insert directly into your worksheet in the same way that Equation Editor is inserted. To open click Insert then Object then OK.

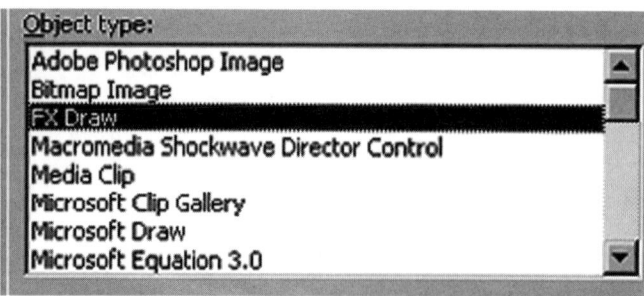

There are many images available, here are some samples.

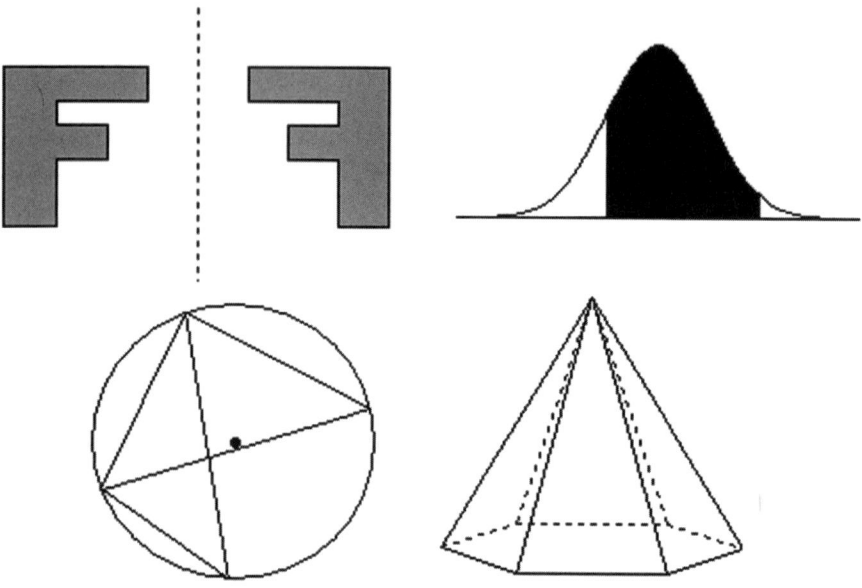

Capturing screen images and including them in your worksheet

Students will find it useful if parts of the software package they are using are shown in their worksheet.

	A	B	C	D
	Positive Integers	Even Numbers	Odd Numbers	Square Numbers
1				
2	1	2	1	
3	=a2+1			
4	Enter a formula in cells A3, B3 and			
5	C3 to generate the first 100			
6	numbers in each of the above 3			
7	series. The formula required for			
8	cell A3 is done for you.			
9				
10	Use your series in column A to			
11	generate the first 100 square			
12	Numbers in Column C. Enter your			
13	formula in cell D2.			
14				

SUM = =a2+1

A screen shot from Excel 97, captured using Paint Shop Pro 5.

You can capture parts of a screen in a variety of ways.

Using the Print Screen button on your keyboard

This method is only recommended if you do not have specific screen capture software. When the Print Screen button is pressed, a copy of the entire screen is sent to the clipboard, where it is held in the computer's memory until you require it.

If you have a Windows operating system then Paint will be one of the programs available. It is normally found in Accessories.

To access Paint, click Start, then Accessories, then Paint.

Once Paint is loaded, click Edit and Paste the screen image from the clipboard.

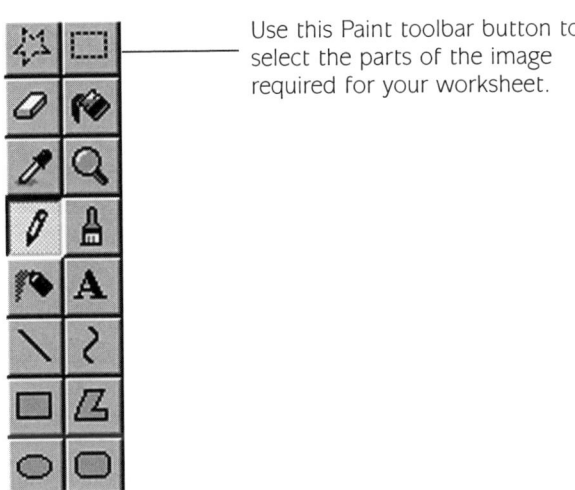 Use this Paint toolbar button to select the parts of the image required for your worksheet.

Early versions of Paint do not have the selection facility. You will have to rub out the parts of the image you do not require!

Once your selection is made, click Edit then Copy and return to your word processing package and Paste your image where you require it in your worksheet.

Using screen capture software

Sophisticated image editing packages like Paint Shop Pro 5 and above and Photo Plus include screen capture features. They are more difficult to use than Paint but they offer many more editing facilities for images.

Alternatively you can purchase screen capture packages like Grabit. These are quick and easy to use.

Some screen capture programs like Grabit may be downloaded from the Internet for evaluation purposes.

Organising your files

If your school is using a network, it is a good idea to set up your own area for the Mathematics department and allow more than one person to log on at the same time. In this area you can create your own Mathematics folders where files should be saved. The Network Manager should arrange for the whole network to be backed up every night, so you do not need to keep separate backups on floppy disk. If your school has the facility to save onto a CD then this is the best option. A CD is more reliable than a floppy disk and can contain the equivalent of about 450 high-density floppy disks.

Similar areas can be set up on a stand-alone PC but remember to keep backups of your files on floppy disks. This can be a lot of disks! Better still, save your files on CD or Zip drives.

It is useful to save your work in folders. This makes specific materials easier to find for future use. In Windows 95 and 98 use Windows Explorer to create Mathematics **folders**. In Windows 3.x use File Manager to create Mathematics **directories**.

Windows Explorer is usually found by clicking the Start button then Programs then Windows Explorer.

When Windows Explorer is open, click File, New and select Folder.

A new folder icon will appear

Rename the new folder Mathematics Area

In the **All Folders** section of Explorer, click on your new Mathematics Area folder to open it.

- Dos
- Encarta
- Exambank ——————————————————————— A closed folder
- Fraction
- Graph   An open folder
- Hpfonts
- Ka
- Kpcms
- Life
- Mathematics Ar
- Norton

With the new folder open, you can create new folders within the main one, to suit your needs.

- Fraction
- Graph
- Hpfonts
- Ka
- Kpcms
- Life
- Mathematics Area
  - Administration ——————————————————— A collection of possible folders to
  - Examinations   keep all your department's files
  - Ks3
  - Ks4
  - Mechanics
  - Performance management
  - Pure
  - Statistics
  - StudentsTest Results
- MathNow

Save every new file in the appropriate folder.

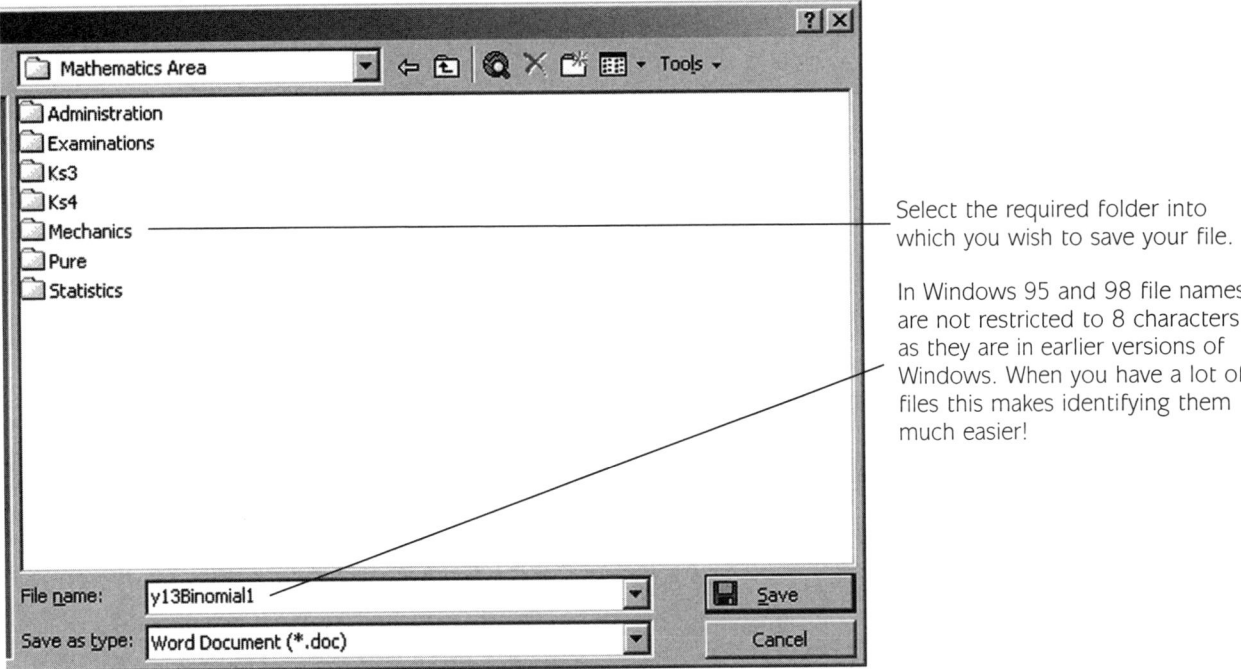

Select the required folder into which you wish to save your file.

In Windows 95 and 98 file names are not restricted to 8 characters as they are in earlier versions of Windows. When you have a lot of files this makes identifying them much easier!

You could make identification of your documents easier by adding text at the top and/or the bottom of the pages. This text is called a header (at the top of the page) and a footer (at the bottom of the page).

To insert a header and/or footer, select View in the main screen menu, then Header and Footer.

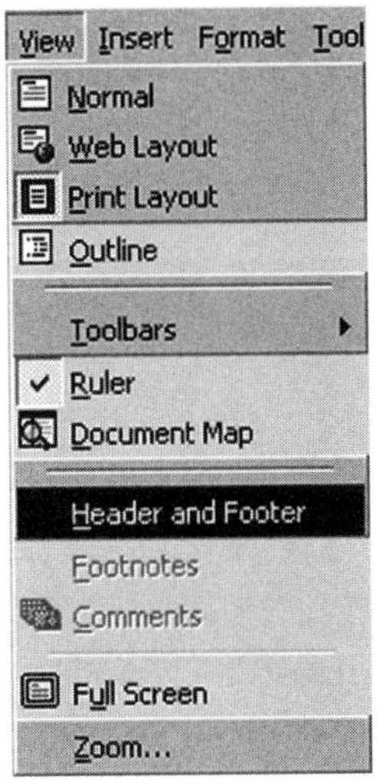

Insert Date: the date automatically changes, if necessary, each time you open the document.

Click this button to switch to the Footer.

You will then see this menu box.

Insert Time

Type the document name in the header or footer and add dates or other details as necessary.

Finally, it is a good idea to keep hard copies of all your worksheets, together with solutions!

A hard copy is a print out of any document.

See the section on Using an Intranet on page 29 if you wish to make files available for students to print off for revision.

Different versions of Word

Microsoft Word software is constantly updated and there are various versions available. Each version offers slightly different options and facilities and it is advisable to familiarise yourself with the version available within your own school. All of the different versions have a similar operating framework, but to help there follows some notable differences.

When Version 6 was released, the software introduced use of the right button on the mouse. If you **right** click on the mouse in Word 6 and above you will get a dialogue box like this.

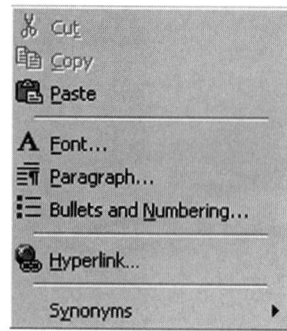

Both Word 6 and Word 97 have multiple Undo and Redo buttons. These buttons allow you to undo or redo previous actions. In Word 2 you can only undo the previous action.

In Word 97 and above, pasted images can be moved easily around the screen. However, in Word 2 and 6 the images need to be pasted into a frame before they can be moved around. To remove the border round a frame in Word 6, click the frame to produce handles then Format, Borders and Shading and select Borders, None. This is similar to Word 2.

To insert a frame in Word 2 and 6, click View, Page Layout, then click on Insert, Frame from the main tool bar.

In Word 97 and above, this dialogue box will appear when you right click inside a drawing shape.

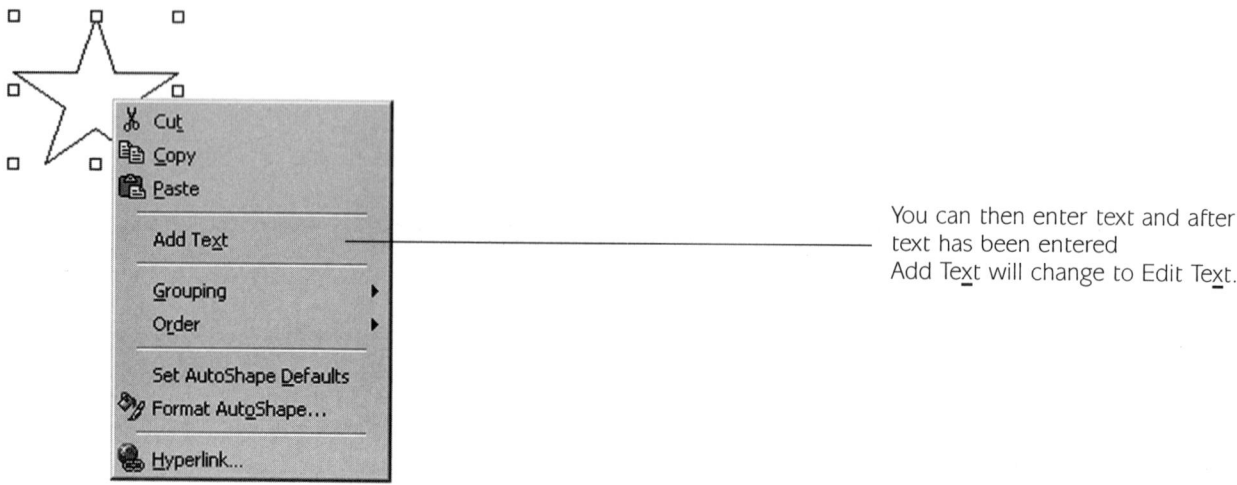

You can then enter text and after text has been entered Add Text will change to Edit Text.

When typing basic text into Worksheets, there is no significant difference between Word 2 and Word 6, but there is no drawing bar in Word 2.

A practical session on using the Equation Editor

When creating worksheets it is more than likely that you will need to incorporate mathematical signs and symbols. To help you develop a working knowledge of Equation Editor try this practical session.

Step 1: Open a new document in Word.

Step 2: Type the title **Lesson 1**

Step 3: Save the document as lesson1.doc

Step 4: Open your version of Equation Editor.
To open the Equation Editor click Insert then Object. The Object box will then appear. Double click on Equation Editor in the Object box.

The diagram below shows the Equation Editor toolbar with some of the buttons needed to insert mathematical signs and symbols.

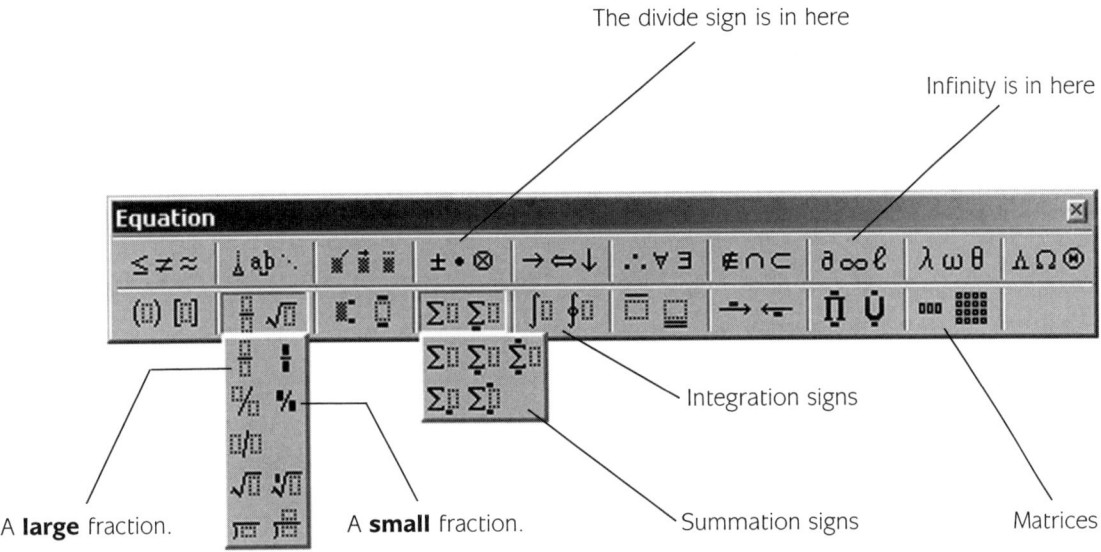

The divide sign is in here

Infinity is in here

Integration signs

A **large** fraction.   A **small** fraction.   Summation signs   Matrices

When using Equation Editor the text will not automatically go onto the next line as it would normally. You must press the Enter key when you need to go onto another line.

Using the hints given, try to recreate the following worksheet.

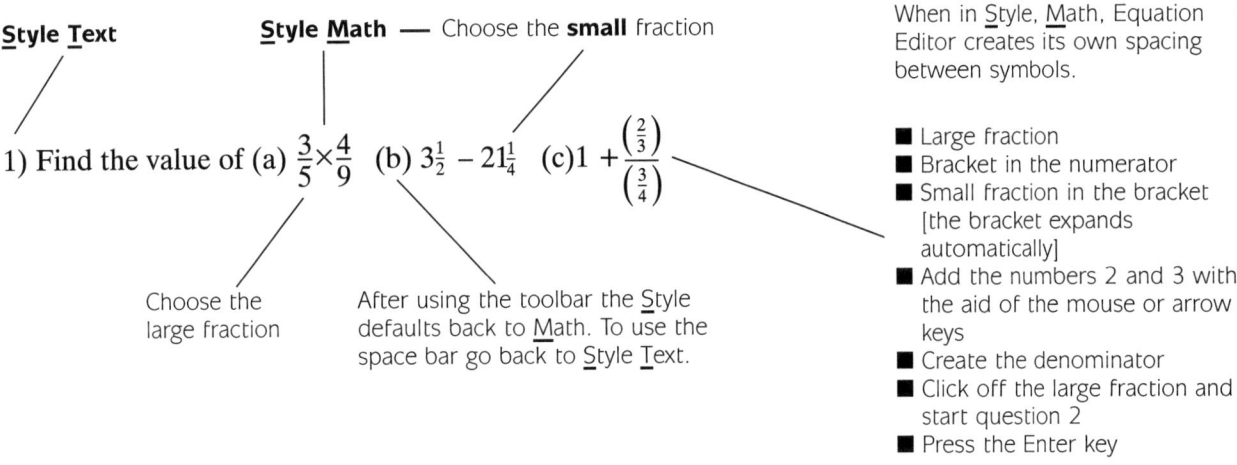

Style Text

Style Math — Choose the **small** fraction

When in Style, Math, Equation Editor creates its own spacing between symbols.

1) Find the value of (a) $\frac{3}{5} \times \frac{4}{9}$ (b) $3\frac{1}{2} - 21\frac{1}{4}$ (c) $1 + \frac{\left(\frac{2}{3}\right)}{\left(\frac{3}{4}\right)}$

Choose the large fraction

After using the toolbar the Style defaults back to Math. To use the space bar go back to Style Text.

■ Large fraction
■ Bracket in the numerator
■ Small fraction in the bracket [the bracket expands automatically]
■ Add the numbers 2 and 3 with the aid of the mouse or arrow keys
■ Create the denominator
■ Click off the large fraction and start question 2
■ Press the Enter key

2) Evaluate 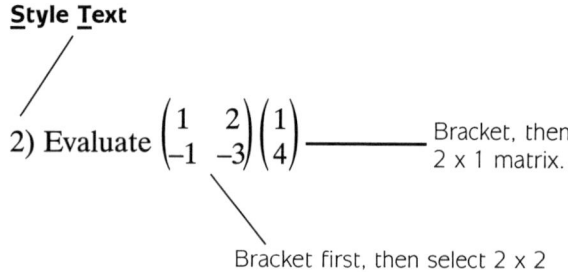 $\begin{pmatrix} 1 & 2 \\ -1 & -3 \end{pmatrix}\begin{pmatrix} 1 \\ 4 \end{pmatrix}$ ————— Bracket, then
2 x 1 matrix.

Bracket first, then select 2 x 2
matrix and enter the values.

3) Evaluate $4^2 + 3^3$

Type 4 then select
superscript and enter 2.

4) If $3^n = (3^4)^3$, find the value of n

Bracket first, 3, superscript, enter 4, move outside the bracket,
superscript, enter 3, click off the superscript level, Style, Text.

5) Simplify $\sqrt{(\sin^2\theta + \cos^2\theta)}$

Choose square root sign, bracket, type sin, superscript, 2.
The bracket will expand automatically.

Continue, with fewer hints!

6) Evaluate $\int_{0}^{\frac{8}{3}} \tan^2\theta \, d\theta$

Select the required Style and signs
and enter the text where needed.

7) Evaluate $\sum_{r=0}^{\infty} \left(\frac{1}{2}\right)^r$

Select the required Style and signs
and enter the text where needed.

Style Math

8) Using the quadratic formula $x_{1,2} = \dfrac{-b \pm \sqrt{b^2 - 4ac}}{2a}$, solve the
quadratic equation $3x^2 - 4x - 6 = 0$, giving your answer to 3
decimal places.

9) Evaluate $2^{3^{3^3}}$

10) Sketch the curve $y = \left|\dfrac{1}{\cos x}\right|$, for $\frac{-\pi}{2} \le x \le \frac{\pi}{2}$

When you have finished using Equation Editor,
click out of the window to return to your document
and remember to save it again!

Projecting your software onto a screen

Demonstrating a mathematical concept with a visual display is a very powerful way to aid a student's understanding of a topic. An animated image is worth more than a thousand words.

Graphing packages are ideal for this, as is a demonstration of how a software package, like Excel, works.

Linear and quadratic transformations, calculus and geometric transformations are all topics which can be introduced with this method of teaching.

Using this technique to introduce the function e^x

The function e^x is defined as being a function whose gradient function has the same value, at every value of x, as the function itself. The value of e is approximately 2.71828...

The software package Omnigraph has a facility to demonstrate this connection between a function and its derivative.

The diagram below shows the curve $y = a^x$, where a has initially been

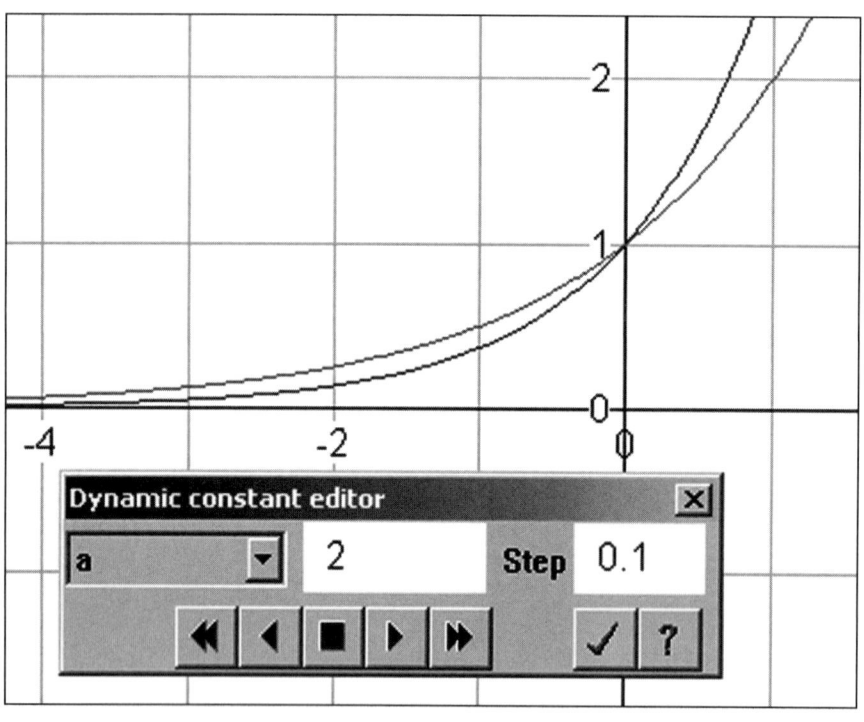

Data Projectors are a must if your school can afford them. They are expensive and the more expensive, the better they are. But they are worth the investment. The greater number of lumens the bulb has, the clearer the image is on screen … and the more expensive the bulb!

As a whole school investment they will prove invaluable for teaching purposes, open evenings, courses, visiting speakers, or anyone using Power Point – a Microsoft package enabling the production of electronic slide shows.

Omnigraph for Windows is now better than ever. Version 2 enables the user to enter a generic function like $y = a^x$ and watch how the function and its derivative change as a is changed.

Note:
Version 1 does not have this facility and Version 2 will only run on Windows 95 or above.

Notes on how to use the Dynamic constant editor can be found on page 15.

These three screen shots demonstrate how the gradient function of $y = a^x$ moves closer to the function itself as a moves from 2.4 to 2.7 in steps of 0.1

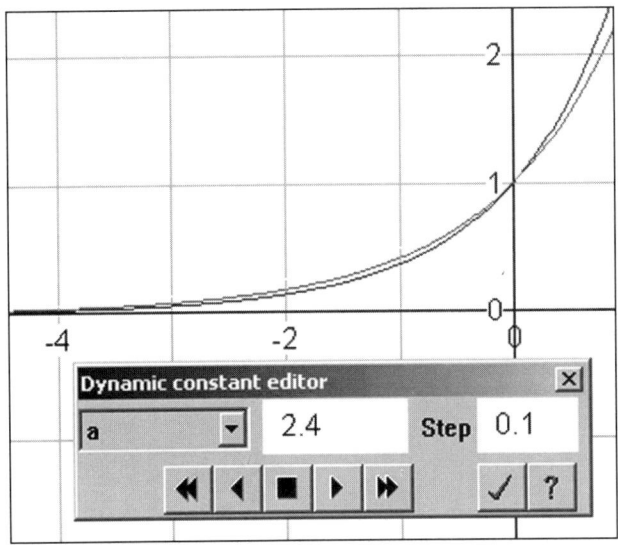

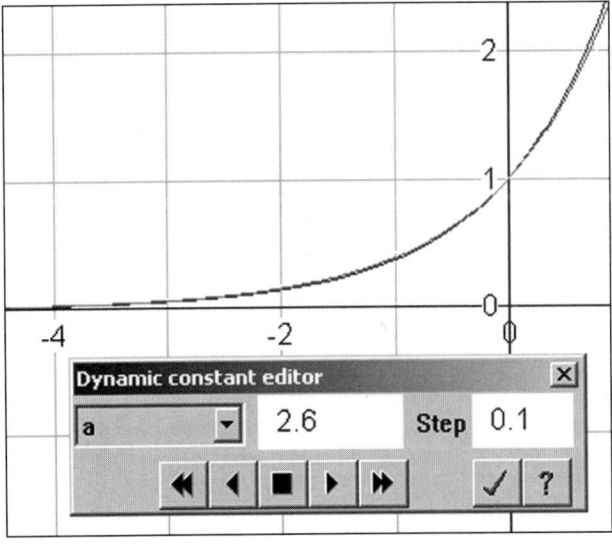

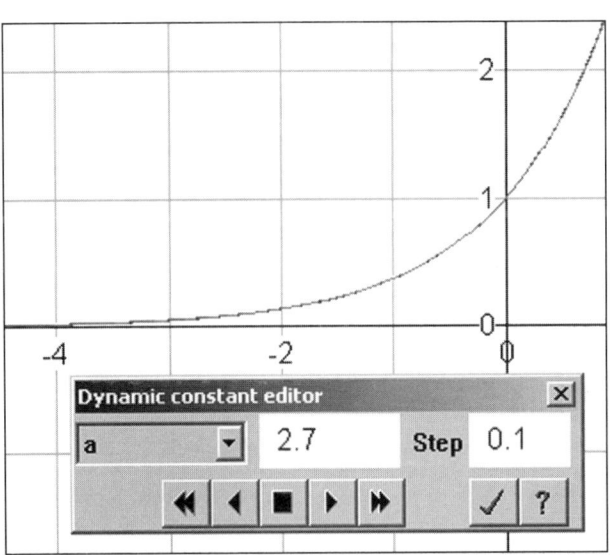

You can introduce trigonometric functions and their gradient functions in a similar way:

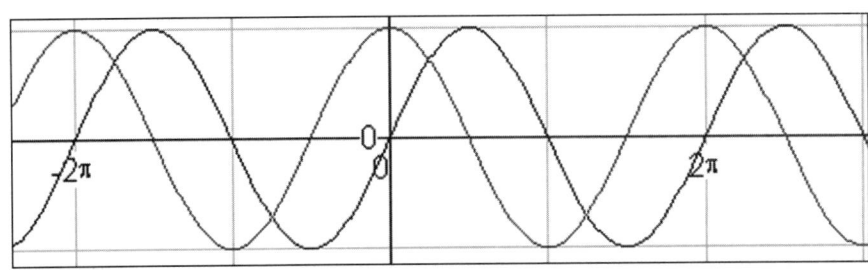

$y = \sin x$ and its gradient function $\cos x$

This graph demonstrates that the derivative of $y = \cos 4x$ is the function $-4\sin 4x$

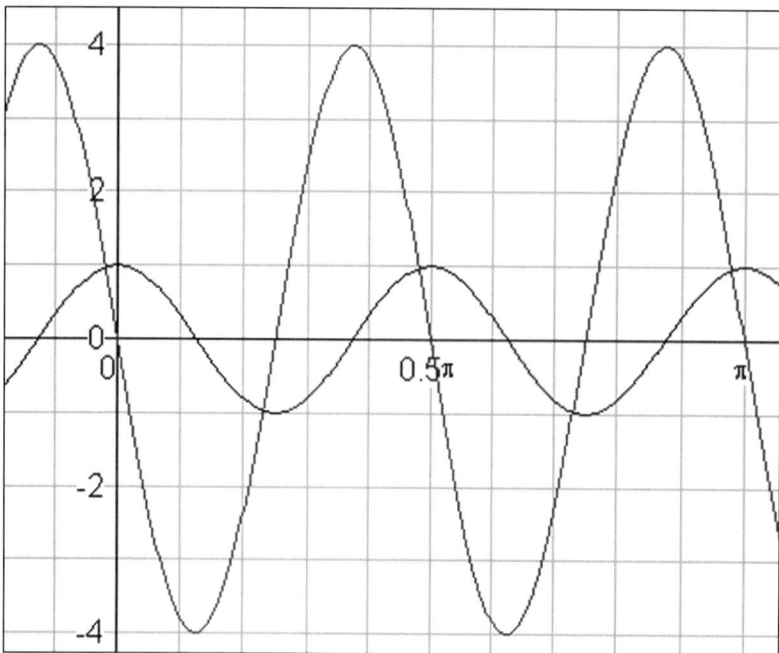

An understanding of function transformations can be enhanced in this way.

As an alternative to this method of whole class display you can provide a worksheet so that individual students can investigate Mathematics at their own pace at their own workstation.

The file onedice.xls on the CD-ROM provides plenty of material for class discussion when run via a data projector. All displays are further enhanced when the computer window is projected onto an *interactive white board.*

Using an Interactive Whiteboard

An interactive whiteboard allows you to use the scroll bars and the menus of the software you are displaying. This is simply done by using your finger 'as the mouse'. It is particularly useful as it saves you from moving away from the whiteboard to the PC to use the mouse in the normal way. The board provides much more flexibility with your classroom management as it allows you to remain facing your audience as you work.

You will still need to input any new data at your keyboard.

The whiteboard is even further enhanced if you can purchase additional software designed to allow you to save windows at the whiteboard, which you can then load again during the lesson.

Using these boards you can:
• write on screen using an electronic pen to produce a slide
• bring up a new slide for the next part of the lesson
• turn the board into graph paper, isometric paper etc.
• review the lesson in the plenary by bringing up earlier slides
• save the slides as a presentation to print off for any absent students
• change the pen colour
• turn the pen into a highlighter to identify important points.

The software also allows you to enter data into a software package such as Excel while remaining at the board. It will provide you with an on-screen keyboard.

The boards are expensive and require a data projector fixed to the ceiling to gain the best use, but after 10 minutes with one of these boards you will not want to use anything else.

3 | Teaching Notes

General Use of the ICT Room

Here are some tips to remember when using the ICT suite:

- Liase with your ICT department to familiarise yourself with the ICT skills of your students.

- If you need to book the ICT room(s) do so well in advance.

- If you require specific files or equipment on the network ensure you give them to your technician in good time.

- Ensure the students are given clear specific targets.

- Ensure the students know which software package you wish them to use and that they know where to find it.

- Provide clear instructions on the board, via a handout or via the intranet, so that students can work independently whilst you talk to individuals.

- When talking to the whole class, ask them to switch off their computer screens. This stops pupils from being distracted.

- Ensure the pupils are clear about what you want them to print. Make sure they add their name to whatever they are printing. Handing back hard copies yourself enables you to monitor progress and stops pupils waiting by the printer. Colour printing is generally expensive and slow and should not be necessary in mathematics lessons. See page 13 for some hints on printing from Excel, page 20 to see how to add text in MSWLogo and page 16 to see how to add text in Omnigraph.

- Allow time at the end of the lesson for pupils to save their work, exit all programs and log off from the computer.

- Allow time for a plenary session.

- Make sure the pupils leave the rooms and equipment ready for the next class.

- See pages 24–28 for hints on using the internet.

Using the activity sheets: ICT and the three-part lesson

Structuring your ICT lessons will help make sure you get the most out of your time in the computer room. The example lesson given below is based around Activity sheet SSM14:

- **5–10 minute introduction**
  Start by outlining the content and aims of the lesson:
    "In this lesson you will look at the connection between straight lines and transformations."

You can then briefly discuss the different types of transformation that will be used and let pupils know where they can find all the files they need to complete the activity.

- **35–40 minute computer work**
  Pupils should be encouraged to keep a record of their results and be reminded to save their work regularly in case of a computer failure. During this interactive period, time should be spent with individual pupils discussing the activity and reminding them to record their results.

- **10 minute plenary session**
  Always leave time at the end of the session to discuss the activity.
    "Describe the connection between the lines $y = x$ and $y = x - 3$."
    "Give two coordinates on the line $y = 2x$."
    "Describe the transformations to change $y = x$ to $y = 3x - 2$."

The activity sheets can be used as follows:

- printed out for the pupils to use and write on as appropriate. For activities where the pupils are accessing websites, this has the disadvantage that the pupils have to enter the website address, accurately

- as a networked file that can be read on screen. Adobe Acrobat Reader will need to be installed on each machine for this (it can be downloaded free of charge). The advantage of this is that there are hyperlinks to all files used and websites. The pupil just clicks on the file name or website address and they are taken automatically to the file or website.

> Adobe Acrobat Reader can be downloaded free from www.adobe.com/products/acrobat/readstep.html/.

Websites

Links to appropriate websites are given in some worksheets. Although these were up to date at the time of writing, it is ESSENTIAL for teachers to preview these sites before using them with pupils. This will ensure that the web address (URL) is still accurate and the content is suitable for your needs.

We suggest that you bookmark useful sites and consider enabling pupils to access them through the school intranet. We are bringing this to your attention as we are aware of legitimate sites being appropriated illegally by people wanting to distribute unsuitable or offensive material. We strongly advise you to purchase suitable screening software so that pupils are protected from unsuitable sites and their material.

If you do find that the links given no longer work, or the content is now unsuitable, please let us know. Details of changes will then be posted on our website. You can also check our website for any changes that we know of.

Using the ICT Activities for Mathematics CD-ROM

Using your CD-ROM

The ICT Activities for Maths CD-ROM contains:

- all the activity sheets in Portable Document File (PDF) format. These can be opened in Adobe Acrobat Reader and viewed on screen or printed out.

- links to websites so that you can download the 30-day evaluation versions of Omnigraph and The Geometer's Sketchpad, and the full version of MSWLogo.

- Excel spreadsheet files to use with some worksheets. The Excel spreadsheets use macros so make sure your computer security settings allow this (or just click **Enable Macros** when opening the spreadsheets).

   In the spreadsheet files, a pupil can only edit cells which he/she needs. All other cells are protected.

Running the CD-ROM

The disk should run automatically when you put it in your CD drive. If it doesn't, double-click on your CD drive (usually E:) under **My Computer**. When the CD opens you will see the main menu selection screen. From here you can access the PDF files of the worksheets, the software demos and the Excel spreadsheet files.

Networking the CD-ROM

You can download the whole CD-ROM onto a server to access the files from more than one computer at the same time. This will allow your students to work on the same worksheet or spreadsheet on screen at the same time.

Simply drag the folder 'ICT Activities Network Version' from the CD onto your C: drive. The CD-ROM will run by double clicking ICT.exe from this folder.

Editing spreadsheets and worksheets

If you have a full version of Adobe Acrobat you can edit the worksheets to create customised lessons and activities.

You can edit the Excel spreadsheets in a normal version of Microsoft Excel. The password for editing the spreadsheets is 'galois'.

CD-ROM technical support

The minimum specification needed to run the ICT Activities for Mathematics CD-ROM is
- Pentium class PC
- Windows 95/98, NT or ME
- 32 Mb RAM
- CD-ROM drive
- keyboard, mouse.

If you experience any problems with the CD-ROM, please phone our technical support line on 01865 888108.

Selecting a worksheet

The following tables will help you select an appropriate worksheet for your pupils. The worksheets are in three main groups:

• Basic Excel skills

• KS3

• KS4.

Within each key stage, the worksheets are grouped by NC attainment target (Ma2, Ma3 and Ma4). Within each table, the worksheets increase in difficulty as you go down the table.

Each table has the following:

• sheet and title

• description: main objectives of worksheet

• requirements: what software package, files and resources are required for the worksheet

• assumed knowledge of package: the skills (for the software package) that pupils require to get the most out of the worksheet

• support: suggested worksheets for you to give pupils whose assumed knowledge is weak

• National Curriculum reference: in the KS4 tables, references refer to both the Foundation and Higher programmes of study unless indicated by an F or H in brackets after each reference.

In addition, there is a table which allows you to select worksheets by National Curriculum reference.

Excel					
Sheet and title	*Description*	*Requirements*	*Assumed knowledge of package*	*Support*	*NC ref.*
EX1 Getting started in Excel 1	Formatting and sorting numbers and text in Excel.	Excel: student1.xls	Can use a mouse and keyboard.	–	–
EX2 Getting started in Excel 2	Entering and using formulae to generate sequences in Excel.	Excel: student2.xls	Can use a mouse and keyboard.	–	ICT 2d
EX3 Getting started in Excel 3	Using Chart Wizard in Excel to create bar charts, pie charts and line graphs.	Excel: student3.xls internet	Can use a mouse and keyboard.	–	ICT 3b

Sheet and title	Description	Requirements	Assumed knowledge of package	Support	NC ref.
NA1 My new room	Using a spreadsheet to model varying costs and adjusting them to a target cost.	Excel: mynewroom.xls	Edit data on a given Excel template. Enter formulae in Excel to look at percentage change.	EX2	1a, 2e, 3e
NA2 Unit conversions	Using a spreadsheet to convert metric and imperial units. Using a given website site to convert currency.	Excel internet	Create formulae in Excel to study unit conversions. Create line graphs in Excel.	EX2 EX3	4a, 5f
NA3 A trail of primes	Using internet sites to gather information on prime numbers.	internet	None.	–	2a, Ma1
NA4 Investigating decimal fractions	Using a calculator and spreadsheet to work out decimal equivalents of rational numbers, including recurring cycles and digit patterns. Searching for irrational numbers on the internet.	calculator Excel: decimal.xls internet	None.	–	2d
NA5 Running an activity weekend	Using an existing spreadsheet file to model money.	Excel: activityweekend.xls	Edit and interpret data in Excel.	EX2	1a, 1d
NA6 Create your own modelling project	Using a spreadsheet to model costs of three people buying a mobile phone. For more able pupils.	internet Excel	Can design own spreadsheet model.	EX2	1a, 1d
NA7 Personal finances and percentages	Using a given internet site to look at mortgages and a given spreadsheet file to model percentage changes.	internet Excel: house.xls	Enter formulae into a spreadsheet.	EX2	2e, 3e
NA8 Money	Using a spreadsheet for modelling, including work on VAT and bar charts.	Excel: starters.xls	Edit a given Excel template. Use formulae to add VAT and sum data.	EX2	2e, 3e, 3m

Sheet and title	Description	Requirements	Assumed knowledge of package	Support	NC ref.
SSM1 Plotting coordinates 1	Using a graphing package to draw polygons.	Omnigraph	Entering coordinates in Omnigraph.	–	1a, 2g, 3e
SSM2 Plotting coordinates 2	Using a graphing package to create polygons and lines of symmetry.	Omnigraph access to a printer	Create polygons in Omnigraph.	SSM1	1a, 1d, 1e, 2g, 3e
SSM3 Investigating angles 1	Using a dynamic geometry package to investigate angles on a straight line and angles round a point.	The Geometer's Sketchpad	None.	–	2a
SSM4 Estimating angles	Using a dynamic geometry package to estimate and check the size of acute, obtuse and reflex angles.	The Geometer's Sketchpad	Draw lines and measure angles in The Geometer's Sketchpad.	SSM3	2b
SSM5 Drawing angles and lines	Using Logo to draw acute, obtuse and reflex angles, parallel and perpendicular lines, equilateral triangles, squares and parallelograms.	MSWLogo	Use simple MSWLogo commands to create acute, obtuse and reflex angles, parallel and perpendicular lines and some related polygons.	–	2b, 2c, 2f, 4j
SSM6 Investigating angles 2	Using a dynamic geometry package to investigate corresponding and alternate angles and parallelograms.	The Geometer's Sketchpad	Draw lines and measure angles in The Geometer's Sketchpad	SSM3	2c, 2f, 2g
SSM7 Investigating angles in triangles and quadrilaterals	Using a dynamic geometry package to investigate the angles in triangles and quadrilaterals.	The Geometer's Sketchpad	Create lines, triangles and quadrilaterals in The Geometer's Sketchpad.	SSM6	2c, 2d, 2f, 2g
SSM8 Imagine	Using Logo and the internet to investigate infinity by looking at area and perimeter.	MSWLogo: infinity.lgo internet	Enter commands into a given MSWLogo file.	SSM5	2e, 4f, 4j (also Ma1)
SSM9 Coordinates, translations, reflections and rotations	Using a graphing package to look at translations, reflections and rotations.	Omnigraph	Enter coordinates to create shapes and apply transformations in Omnigraph.	SSM2	3a, 3b, 3e
SSM10 Coordinates, enlargements and ratios	Using a graphing package to look at enlargements and ratios.	Omnigraph	Enter coordinates to create shapes and apply transformations in Omnigraph.	SSM2	3a, 3c, 3d, 3e

Sheet and title	Description	Requirements	Assumed knowledge of package	Support	NC ref.
SSM11 Coordinates and straight lines	Using a graphing package to look at the equations of horizontal and vertical lines.	Omnigraph	Enter coordinates in Omnigraph.	SSM1	1a, 3e (also Ma2 6e)
SSM12 Patterns	Using a graphing package to create unusual symmetrical shapes using common trigonometric functions. Introduces polar coordinates. Looking at other curves at a given website.	Omnigraph internet	Enter a variety of polar equations, mainly trigonometric and involving fractions, into Omnigraph.	SSM2	1d, 1f, 1g, 1i, 1j, 4j
SSM13 Investigating straight lines 1	Using a dynamic geometry package to investigate the gradients of parallel and perpendicular lines to understand the line $y = mx + c$.	The Geometer's Sketchpad	Create lines, measure slope and add axes in The Geometer's Sketchpad.	SSM3	2a, 2c (also Ma2 6e, 6g, 6h)
SSM14 Investigating straight lines 2	Using a graphing package to produce straight lines by a series of transformations – change $y = x$ into $y = mx + c$.	Omnigraph	Enter linear equations, sometimes involving fractions, into Omnigraph.	SSM9	3a, 3b, 4j
SSM15 Drawing bearings	Using Logo to draw the points of the compass and bearings.	MSWLogo	Use simple MSWLogo commands.	SSM5	4b, 4j

Sheet and title	Description	Requirements	Assumed knowledge of package	Support	NC ref.

Key Stage 3: Ma4

Sheet and title	Description	Requirements	Assumed knowledge of package	Support	NC ref.
HD1 Mathematics through the ages	Using a given website to find information about famous mathematicians to help solve a crossword puzzle.	internet	None.	–	1a, 1b, 1c, 1d, 1g, 4b
HD2 A day out in London	Using the internet to find out information for a day out and makes use of the 24-hour clock and the calendar. Extension: capture images and produce a slide show of their day. Use a spreadsheet to record expenditure.	internet Excel	Enter data and formulae in Excel.	EX2	1a, 1b, 1c, 1d, 3b, 5a, 5e
HD3 Flights	Using information from a given website to find out flight times. Makes use of the 12- and 24-hour clock and the calendar.	internet	None.	–	1a, 1d, 3b, 5a, 5e
HD4 Looking at data 1	Using a spreadsheet including creating formulae, mean averages, sorting, analysing data, scatter graphs and correlation. Modelling changes to league positions by changing the number of points for a win.	Excel: Soccer League Tables.xls	Enter formula into Excel. Sort data, produce scatter graphs and look for correlation.	EX2 EX3	1f, 3b, 4a, 4e, 4j, 5f
HD5 Looking at data 2	Using a spreadsheet to describe social trends, interpret data, work out percentages and make predictions.	Excel: olympics.xls access to a printer	Produce scatter graphs and drag data in Excel. Enter a formula to calculate percentages.	EX2 EX3	1f, 5b, 5c, 5d, 5f
HD6 Business trip	Using the internet to plan a business trip. Makes use of the 24-hour clock. Use given sites to do currency conversions and calculate distances travelled.	internet	None.	–	1a, 1d, 3b, 5a, 5e (also Ma2 3o, 3q, 4a)
HD7 Cumulative frequency	Using a spreadsheet to estimate medians, upper and lower quartiles and interquartile ranges from cumulative frequency curves.	Excel access to a printer	Enter formulae into Excel. Produce a cumulative frequency curve with the aid of a scatter graph in Excel.	EX2 EX3	4a, 4b, 5b

Sheet and title	Description	Requirements	Assumed knowledge of package	Support	NC ref.
HD8 Quartiles, averages and ranges	Using a spreadsheet containing house price data to calculate lower and upper quartiles, interquartile ranges, means and medians. Interpret data.	Excel: houseprice.xls	Enter formulae into Excel.	EX2	1a (iii), 4a, 4b, 5d
HD9 Looking at data 3	Using a spreadsheet to analyse data on transplant procedures and teenage pregnancies, produce scatter graphs, look at trends and lines of best fit.	Excel: transplants.xls access to a printer	Produce scatter graphs in Excel.	EX3	4a, 5b, 5c, 5g
HD10 Probability models	Using a given spreadsheet to compare experimental probabilities and theoretical probabilities.	Excel: onedice.xls, twodice.xls, lottery.xls	None.	–	5a, 5c, 5h, 5j

Key Stage 4: Ma2

Sheet and title	Description	Requirements	Assumed knowledge of package	Support	NC ref. **
NA9 Analysing food	Using a given website to search for recipes. Use the site for unit conversions, ratio and proportions.	internet Excel	Copy data from the internet and paste it into a word processing package. Edit data to reflect different proportion changes to ingredients. Create bar [column] charts in Excel. Use simple formulae in Excel.	EX2 EX3	2f, 4a, 5f(F), 5g(H), (also Ma4, 1a)
NA10 Investigating quadratics 1	Using a graphing package to show how the values of the integers a, b and c affect the shape of the curve $y = ax^2 + bx + c$.	Omnigraph	Enter quadratic equations into Omnigraph.	SSM14	5c(F), 5k(H), 6e(H) (also Ma1)
NA11 Investigating sequences	Using a spreadsheet to generate the Fibonacci sequence and the Golden Number. Look at a given website for information related to Fibonacci numbers.	Excel internet	Enter data and complicated formulae into Excel. Produce line graphs in Excel.	EX2 EX3	6a

**References refer to both Foundation and Higher Programmes of Study unless indicated by (F) for Foundation and (H) for Higher.

Sheet and title	Description	Requirements	Assumed knowledge of package	Support	NC ref. **
NA12 Linear functions and their inverse	Using a graphing package to help pupils find the inverse of linear functions.	Omnigraph	Enter linear equations, some containing fractions, into Omnigraph.	SSM14	6b, 6c, 6d(H), 6e(F)
NA13 Investigating quadratics 2	Using a graphing package to produce quadratic curves by completing the square and using translations.	Omnigraph	Enter quadratic equations into Omnigraph.	SSM14	5k(H), 6e
NA14 Approximating a value for π *	Using a spreadsheet and the internet to investigate infinite series, sequences, convergence, percentage errors and other types of numbers. Research given sites for information related to π and irrational, transcendental and imaginary numbers.	Excel internet	Enter formulae into Excel.	EX2	1d, 2a, 2b, 2c, 2d, 6a(H) (also Ma1)
NA15 Infinite series*	Using a spreadsheet to investigate the divergence and convergence of infinite series.	Excel	Enter formulae into Excel.	EX2	6a(H) (also Ma1)
NA16 Solving equations	Using a graphing package to solve quadratic and cubic equations.	Omnigraph	Enter cubic equations into Omnigraph.	NA13	5k(H), 5l(H), 6f(H) but ext. to cubics
NA17 The approximate solution to equations	Using a spreadsheet to solve equations by iteration.	Excel	Enter complicated formulae in Excel.	NA11	5f(F), 5g(H), 5m(H)
NA18 Odd and even functions	Using a graphing package to investigate odd and even functions.	Omnigraph	Enter a variety of equations into Omnigraph. Use trigonometric axes.	NA13	1j, 6f(H) (also Ma1)
NA19 Function transformations	Using a graphing package to investigate function transformations.	Omnigraph	Enter a variety of equations into Omnigraph. Use trigonometric axes.	NA13	6f(H), 6g(H)

*For the more able pupil.

**References refer to both Foundation and Higher Programmes of Study unless indicated by (F) for Foundation and (H) for Higher.

Key Stage 4: Ma3					
Sheet and title	*Description*	*Requirements*	*Assumed knowledge of package*	*Support*	*NC ref.* **
SSM15 Drawing bearings	Using Logo to draw the points of the compass and given bearings	MSWLogo	Use simple MSWLogo commands.	SSM5	4a(H), 4b(F), 4e(H), 4j(F)
SSM16 Investigating circles 1	Using a dynamic geometry package to investigate the tangents and chords of a circle.	The Geometer's Sketchpad	Draw lines and measure angles in The Geometer's Sketchpad.	SSM3	2h(H), 2i(F)
SSM17 Investigating circles 2	Using a dynamic geometry package to investigate properties of circles.	The Geometer's Sketchpad	Draw lines and measure angles in The Geometer's Sketchpad.	SSM16	2h(H), 2i(F)
SSM18 Triangles, bisectors and circles	Using a dynamic geometry package to look at the different circles obtained when the angles and sides of a triangle are bisected.	The Geometer's Sketchpad	Draw lines and add points, angle bisectors and perpendicular bisectors in The Geometer's Sketchpad.	SSM17	4b(H), 4c(H), 4j(F)
SSM19 Investigating geometrical sequences*	Using a spreadsheet to investigate the limit of a geometric sequence of circles and polygons. Uses trigonometry, angles in polygons in both degrees and radians and limits and formula creation.	calculator Excel	Knowledge of trigonometry and geometry. Enter trigonometric formulae in Excel. An understanding of radians. 3D trigonometry (for the extension).	EX2	2f(H), 2g(H), 2h(H) (also Ma1)
SSM20 Approximating a value for π*	Using Pythagoras' theorem and area (sum of areas of trapeziums and triangle) to estimate a value for π using rough calculations and a spreadsheet.	Excel scientific calculator	A good knowledge of a spreadsheet.	EX2	2f(H), 2h, 4d(H) (also Ma1)
SSM21 Investigating trigonometric curves	Using a graphing package to investigate the connections between trigonometric curves, making use of transformations.	Omnigraph	Enter trigonometric equations into Omnigraph.	NA16	3a, 3b (also Ma1 and Ma2 6g(H))

*For the more able pupil.
**References refer to both Foundation and Higher Programmes of Study unless indicated by (F) for Foundation and (H) for Higher.

Sheet and title	Description	Requirements	Assumed knowledge of package	Support	NC ref. **
HD11 Looking at data 4	Using a spreadsheet to look at totals, averages (mean, median and mode), scatter graphs, correlation. Extension: entering and interpreting data acquired from given websites.	Excel: euro.xls internet	Enter formulae into a given Excel file. Produce scatter graphs in Excel.	EX2 EX3	3a, 3b(H), 3c(F), 4a, 4e(H), 4g(F), 4j(H), 5f
HD8 Quartiles, averages and ranges	Using a spreadsheet containing house price data to calculate lower and upper quartiles, interquartile ranges, means and medians. Interpret data.	Excel: houseprice.xls	Enter formulae into Excel.	EX2	4a, 4b(F), 4e(H), 4j(H), 5d
HD12 Estimating the mean	Using a spreadsheet to estimate the mean from a grouped frequency table.	Excel	Enter formulae into Excel.	EX2	4e(H), 4g(F), 4j(H)
HD10 Probability models	Using a given spreadsheet to compare experimental probabilities and theoretical probabilities.	Excel: onedice.xls, twodice.xls, lottery.xls	None.	–	5h
HD13 Investigating standard deviation	Using a spreadsheet to calculate the standard deviation of sets of data.	Excel	Enter formulae into Excel.	EX2	Not required for GCSE Maths starting in Sep 2001.

**References refer to both Foundation and Higher Programmes of Study unless indicated by (F) for Foundation and (H) for Higher.

Finding a worksheet by National Curriculum reference

NC Ref.	Worksheets
Key stage 3	
Ma2, 1a	NA1, NA5, NA6
1d	NA5, NA6
2a	NA3
2d	NA4
2e	NA1, NA7, NA8
3e	NA1, NA7, NA8
3m	NA8
3o	(HD6)
3g	(HD6)
4a	NA2, (HD6)
5f	NA2
6e	(SSM11), (SSM13)
6g	(SSM13)
6h	(SSM13)
Ma3, 1a	SSM1, SSM2, SSM11
1d	SSM2
1e	SSM2
1d	SSM12
1f	SSM12
1g	SSM12
1i	SSM12
1j	SSM12
2a	SSM3, SSM13
2b	SSM4, SSM5
2c	SSM5, SSM6, SSM7, SSM13
2d	SSM7
2e	SSM8
2f	SSM5, SSM6, SSM7
2g	SSM1, SSM2, SSM6, SSM7
3a	SSM9, SSM10, SSM14
3b	SSM9, SSM14
3c	SSM10
3d	SSM10
3e	SSM9, SSM10, SSM11
3e	SSM1, SSM2
4b	SSM15
4f	SSM8
4j	SSM5, SSM8, SSM12, SSM14, SSM15

NC Ref.	Worksheets
Ma4 1a	HD1, HD2, HD3, HD6, HD8
1b	HD1, HD2
1c	HD1, HD2
1d	HD1, HD2, HD3, HD6
1f	HD4, HD5
1g	HD1
3b	HD2, HD3, HD4, HD6
4a	HD4, HD7, HD8, HD9
4b	HD1, HD7, HD8
4e	HD4
4j	HD4
5a	HD2, HD3, HD6, HD10
5b	HD5, HD7, HD9
5c	HD5, HD9, HD10
5d	HD5, HD8
5e	HD2, HD3, HD6
5f	HD4, HD5
5g	HD9
5h	HD10
5j	HD10

Key stage 4: Foundation	
Ma2 1d	NA14
1j	NA18
2a	NA14
2b	NA14
2c	NA14
2d	NA14
2f	NA9
4a	NA9
5c	NA 10
5f	NA9, NA17
6a	NA11
6b	NA12
6c	NA12
6e	NA10, NA13
Ma3 2i	SSM16, SSM17
3a	SSM21
3b	SSM21
4b	SSM15
4h	SSM16
4j	SSM18
Ma4 1a	(NA9)
3a	HD11
3c	HD11
4a	HD8, HD11
4b	HD8
4g	HD11, HD12
5d	HD8
5f	HD11
5h	HD13

Key stage 4: Higher	
Ma2 1d	NA14
1j	NA18
2a	NA14
2b	NA14
2c	NA14
2d	NA14
2f	NA9
4a	NA9
5g	NA9, NA17
5k	NA10, NA13, NA16
5l	NA16
5m	NA17
6a	NA11, NA14, NA15
6b	NA12
6c	NA12
6d	NA12
6e	NA10, NA13
6f	NA16, NA18, NA19
6g	NA19 (SSM21)
Ma3 2f	SSM19, SSM20
2g	SSM19
2h	SSM16, SSM17, SSM19, SSM20
3a	SSM20
3b	SSM21
4a	SSM15
4b	SSM18
4c	SSM18
4d	SSM20
4e	SSM15
Ma4 1a	(NA9)
3a	HD11
3b	HD11
4a	HD8, HD11
4e	HD8, HD11, HD12
4j	HD8, HD11, HD12
5d	HD8
5f	HD11
5h	HD13

Teaching notes and answers for the activity sheets

The teaching notes are arranged as follows:

Activity sheet number and title

Requirements: gives details of the package used and any other resources needed such as access to a printer.

File: lists any files that will be needed from the CD-ROM for the activities.

Aims: specific to the sheet.

NC Reference: to the relevant programme of study.

Teacher preparation: what you need to do in preparation for the activity.

Assumed knowledge: gives details of the assumed knowledge of the package that pupils will need to do the activities.

Support: suggests suitable activity sheets for you to use with your pupils if their assumed knowledge of the software is not good enough.

Starting the activity: suggestions for getting the activity started.

Hints: gives any common problems that you or your pupils may come across, as well as helpful hints.

Answers: gives answers for the activity sheet, where appropriate.

EX1: Getting started in Excel 1

Requirements: Excel
File: student1.xls
Aims:
■ format and sort numbers and text.
NC Reference: none
Teacher preparation: check that pupils can access file.
Assumed knowledge:
■ can use a mouse and keyboard.
Support:
Starting the activity:
Hints:
Answers: none.

EX2: Getting started in Excel 2

Requirements: Excel
File: student2.xls
Aims:
■ enter formulae
■ use formulae to generate sequences.
NC Reference: ICT 2d
Teacher preparation: check that pupils can access file.
Assumed knowledge:
■ can use a mouse and keyboard.
Support: EX1
Starting the activity:
Hints: the multiplication sign is *; the division sign is /.
Answers:
column F: in cell F2, $=(F1-1)*3+1$. Terms are: 2, 4, 10, 28, 82, 244, 730, 2188, 6562, 19684, 59050, 177148, 531442, 1594324, 4782970.
column G: in cell G3, $=(G2-G1)*2+G2$. Terms are: 3, 11, 27, 59, 123, 251, 507, 1019, 2043, 4091, 8187, 16379, 32763, 65531, 131067.

EX3: Getting started in Excel 3

Requirements: Excel, internet (extension only)
File: student3.xls
Aims:
■ display data as bar and pie charts, and line graphs
NC Reference: ICT 3b
Teacher preparation: check that pupils can access file; (extension only) that there is access to the internet.
Assumed knowledge:
■ can use a mouse and keyboard.
Support: EX1
Starting the activity:
Hints: bar charts are called column charts in Excel.
Answers: none.

NA1: My new room

Requirements: Excel
File: mynewroom.xls
Aims:
■ use a spreadsheet to model varying costs
■ adjust costs to meet a target.
NC reference: KS3 Ma2 1a, 3e
Teacher preparation: check that pupils can access file; check that pupils can edit the data in the file (i.e. the cells are not protected).
Assumed knowledge:
■ edit data on a given template
■ enter formulae to look at percentage change.
Support: EX1, EX2
Starting the activity:
Hints: use the multiplication factor 0.85 for percentage decrease.

Answers:

1 Over budget by £302.92.

2 Under budget by £585.09.

3 An answer of 8p under budget can be obtained by:

Item	Cost	Quantity	Total
Morning Yellow Stripe	£ 9.99	5	£ 49.95
Convoy Single	£219.00	1	£ 219.00
Eton Athalon 850Mhz	£999.00	1	£ 999.00
Blue Gloss	£ 7.99	1	£ 7.99
Simple Desk Lamp	£ 9.99	1	£ 9.99
Hanging Lamp	£ 34.99	1	£ 34.99
Home Office Desk	£179.00	1	£ 179.00
Total			**£1,499.92**

4 An answer of £1.52 under budget can be obtained by:

Item	Cost	Sale price	Quantity	Total
Morning Yellow Stripe	£ 9.99	£ 8.49	5	£ 42.46
Hideaway Guest Bed	£ 279.00	£237.15	1	£ 237.15
Gateway Laptop	£1,099.00	£934.15	1	£ 934.15
Dreamy Cool Green	£ 10.99	£ 9.34	1	£ 9.34
Adjustable Desk Lamp	£ 24.99	£ 21.24	1	£ 21.24
Candle Stick Lamp	£ 39.99	£ 33.99	1	£ 33.99
Dark Brown Desk	£ 259.00	£220.15	1	£ 220.15
Total				**£1,498.48**

NA2: Unit conversions

Requirements: Excel, internet
File: none
Aims:
■ convert metric to imperial units and vice versa
■ convert Fahrenheit to Celsius and vice versa
■ convert currencies.
NC Reference: KS3 Ma2 4a, 5f
Teacher preparation: check that pupils can access file; that there is access to the internet; website addresses.
Assumed knowledge:
■ create formulae to study unit conversions
■ use a spreadsheet to create line graphs.
Support: EX2, EX3
Starting the activity: demonstrate a conversion graph on the board first.
Hints: if printing, remind pupils to use print preview.
Answers:

1

Kilograms	Pounds	Kilograms	Pounds
0	0	26	57.2
1	2.2	27	59.4
2	4.4	28	61.6
3	6.6	29	63.8
4	8.8	30	66
5	11	31	68.2
6	13.2	32	70.4
7	15.4	33	72.6
8	17.6	34	74.8
9	19.8	35	77
10	22	36	79.2
11	24.2	37	81.4
12	26.4	38	83.6
13	28.6	39	85.8
14	30.8	40	88
15	33	41	90.2
16	35.2	42	92.4
17	37.4	43	94.6
18	39.6	44	96.8
19	41.8	45	99
20	44	46	101.2
21	46.2	47	103.4
22	48.4	48	105.6
23	50.6	49	107.8
24	52.8	50	110
25	55		

(a) 5kg = 11lbs
(b) 33lbs = 15kg
(c) 19kg = 41.8lbs
(d) 4kg = 8.8lbs
(e) 101.2lbs = 46kg
(f) 94.6lbs = 43kg

Kilometres	Miles	Kilometres	Miles
0	0	4.1	2.5625
0.1	0.0625	4.2	2.625
0.2	0.125	4.3	2.6875
0.3	0.1875	4.4	2.75
0.4	0.25	4.5	2.8125
0.5	0.3125	4.6	2.875
0.6	0.375	4.7	2.9375
0.7	0.4375	4.8	3
0.8	0.5	4.9	3.0625
0.9	0.5625	5	3.125
1	0.625	5.1	3.1875
1.1	0.6875	5.2	3.25
1.2	0.75	5.3	3.3125
1.3	0.8125	5.4	3.375
1.4	0.875	5.5	3.4375
1.5	0.9375	5.6	3.5
1.6	1	5.7	3.5625
1.7	1.0625	5.8	3.625
1.8	1.125	5.9	3.6875
1.9	1.1875	6	3.75
2	1.25	6.1	3.8125
2.1	1.3125	6.2	3.875
2.2	1.375	6.3	3.9375
2.3	1.4375	6.4	4
2.4	1.5	6.5	4.0625
2.5	1.5625	6.6	4.125
2.6	1.625	6.7	4.1875
2.7	1.6875	6.8	4.25
2.8	1.75	6.9	4.3125
2.9	1.8125	7	4.375
3	1.875	7.1	4.4375
3.1	1.9375	7.2	4.5
3.2	2	7.3	4.5625
3.3	2.0625	7.4	4.625
3.4	2.125	7.5	4.6875
3.5	2.1875	7.6	4.75
3.6	2.25	7.7	4.8125
3.7	2.3125	7.8	4.875
3.8	2.375	7.9	4.9375
3.9	2.4375	8	5
4	2.5		

(a) 1 km = 0.625 miles
(b) 1 mile = 1.6 km
(c) 4.5 km = 2.8125 miles
(d) 5 miles = 8 km
(e) 6.4 km = 4 miles
(f) 4.25 miles = 6.8 km

Degrees Celsius	Degrees Fahrenheit	Degrees Celsius	Degrees Fahrenheit
0	32	16	60.8
1	33.8	17	62.6
2	35.6	18	64.4
3	37.4	19	66.2
4	39.2	20	68
5	41	21	69.8
6	42.8	22	71.6
7	44.6	23	73.4
8	46.4	24	75.2
9	48.2	25	77
10	50	26	78.8
11	51.8	27	80.6
12	53.6	28	82.4
13	55.4	29	84.2
14	57.2	30	86
15	59		

(a) 10°C = 50°F   (d) 77°F = 25°C
(b) 5°C = 41°F   (e) 30°C = 86°F
(c) 59°F = 15°C   (f) 48.2°F = 9°C

3 (a) kg = lb ÷ 2.2
(b) km = m × 1.6
(c) $°C = \dfrac{°F - 32}{1.8}$

5 (a) The figures change daily – the figures given here are given as an example.

US Dollars	Japanese Yen	UK Pound	Euro
0	0	0	0
1	108.5187195	1	1.6044
5	542.5935974	5	8.0220
10	1085.187195	10	16.0440
15	1627.780792	15	24.0660
20	2170.37439	20	32.0880
25	2712.967987	25	40.1100
30	3255.561584	30	48.1320
35	3798.155182	35	56.1540
40	4340.748779	40	64.1760
45	4883.342377	45	72.1980
50	5425.935974	50	80.2200

(b)

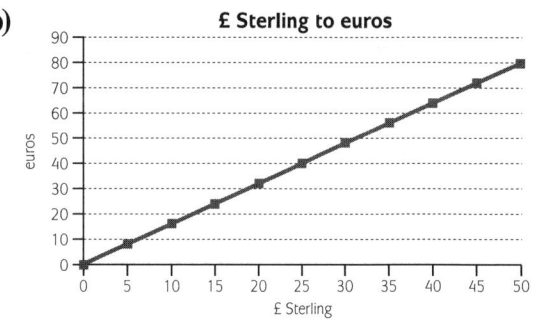

NA3: A trail of primes

Requirements: internet
File: none
Aims:
■ use internet to gather information on prime numbers.
NC Reference: KS3, Ma2 2a, Ma1
Teacher preparation: check that there is access to the internet; website addresses, so that you can see what the pupils will be looking for.
Assumed knowledge:
■ none.
Support:
Starting the activity: discuss how large prime numbers are used in internet security.
Hints: encourage pupils to remain on the given websites during the lesson.
Answers: none

NA4: Investigating decimal fractions

Requirements: calculator, Excel, internet
File: decimal.xls
Aims:
■ use a calculator and spreadsheet to work out decimal equivalent of rational numbers
■ use a spreadsheet to look at recurring cycles and digit patterns
■ searching for irrational numbers on the internet.
NC Reference: KS3 Ma2 2d
Teacher preparation: check that pupils can access the file. Have some spare calculators.
Assumed knowledge:
■ none.
Support:
Starting the activity: explain how any fraction entered into a calculator will produce a terminating decimal like 0.25 or a recurring decimal and how the recurring cycle may not be visible in the calculator display.
Hints: note that the **maximum** recurring cycle a decimal fraction can have is one less than the denominator.
Q9. Worksheet SSM20 can be used to find an estimate for the value of π.
Answers:

4 0.206185567
5 No.
6 The Numbers are: $\frac{1}{19}, \frac{1}{23}, \frac{1}{29}, \frac{1}{47}, \frac{1}{59}, \frac{1}{61}, \frac{1}{109}, \frac{1}{113}, \frac{1}{131}, \frac{1}{149}$
7 $\frac{1}{17389}$ (known to authors)
8 (a) and (b) The same digits occur in each cycle.
9 Yes.

NA5: Running an activity weekend

Requirements: Excel
File: activityweekend.xls
Aims:
■ model money using a spreadsheet.
NC Reference: KS3 Ma2 1a, 1d

Teacher preparation: check that pupils can access the file.
Assumed knowledge:
■ edit and interpret data in Excel.
Support: EX2
Starting the activity: discuss what may be required if organising the finances of an activity weekend.
Hints: if a cell (or cells) has been formatted for currency, you do not need to enter the £ sign. If the £ is entered, you may get an error message.
Answers:

1 New income is £600.
2 With 18 children a cost of £33.33 gives an income of £599.94.
3 Overall profit is £76.19.
5 Overall profit now £56.19.

NA6: Create your own modelling project

Requirements: internet, Excel
File: none
Aims:
■ use a spreadsheet to find which mobile phone will be the best for each of three people.
NC Reference: KS3 Ma2 1a, 1d
Teacher preparation: check that pupils can access the file; there is access to the internet; website addresses.
Assumed knowledge:
■ can design own spreadsheet model (for the more able pupil).
Support:
Starting the activity:
Hints: plan the model on paper first and keep to one type of tariff.
Answers: none

NA7: Personal finances and percentages

Requirements: internet, Excel
File: house.xls
Aims:
■ use a given internet site to find information about mortgages
■ model percentage changes using a spreadsheet.
NC Reference: KS3 Ma2 2e, 3e
Teacher preparation: check that pupils can access the file; there is access to the internet; website addresses; find a region for good property searches.
Assumed knowledge:
■ enter formulae into a spreadsheet.
Support: EX2
Starting the activity: discuss the requirements of buying a house.
Hints: remind pupils to remain on the given website.
Answers:
This example was created on 1st September 2001.
Stefan and Sharmin can borrow £151,250.00.

They decide to buy a property for £128,562.50.
They need to borrow £124,662.50.
On a mortgage with 2-year fixed rate and free valuation the repayments are £811.55.
Their monthly profit/loss is £1,463.45.

NA8: Money

Requirements: Excel
File: starters.xls
Aims:
■ use a spreadsheet for modelling including work on VAT and bar charts.
NC Reference: KS3 Ma2 2e, 3e, 3m
Teacher preparation: check that pupils can access the file.
Assumed knowledge:
■ edit a given Excel file
■ use formulae to add VAT and sum data.
Support: EX2
Starting the activity: discuss how modelling helps in business.
Hints: to multiply numbers in Excel use the * key.
In Excel a bar chart is called a column chart.
Remind pupils that charts are best placed as a new sheet. Excel uses the US date format as its default, i.e. month/day/ year.
Answers:

1 to 5 This spreadsheet shows the number of weekly sales of starters in Ali's restaurant for the week commencing July 10th 2000.

Week Commencing	
07/10/00	£ 841.48
07/17/00	
07/24/00	
07/31/00	

Starter	Cost	Number Sold	Income
Lentil Soup	£1.40	15	£21.00
Mulligatawny Soup	£1.50	12	£18.00
Prawn Cocktail	£2.10	2	£ 4.20
Shish Kebab	£3.95	17	£67.15
Sheek Kebab	£2.25	10	£22.50
Kebab	£2.25	11	£24.75
Lamb Tika	£2.25	16	£36.00
Chicken Tika	£2.25	24	£54.00
Tandoori Chicken	£2.25	45	£101.25
Meat Samosa	£1.80	16	£28.80
Vegetable Samosa	£1.60	15	£24.00
Onion Bhajee	£1.40	32	£44.80
Chicken Chatt	£2.25	9	£20.25
Chana Chatt	£2.10	4	£ 8.40
Prawn Puri	£2.35	11	£25.85
King Prawn Puri	£5.50	9	£49.50
Pakara	£1.90	13	£24.70
Stuffed Pepper	£2.95	7	£20.65
Tandoori Mixed Grill	£2.95	23	£67.85
Vegetable Korma	£2.10	25	£52.50
		316	£716.15 £841.48

6 This spreadsheet shows the number of weekly sales of starters in Ali's restaurant for the week commencing July 17th 2000.

Week Commencing	
07/10/00	£ 841.48
07/17/00	£ 781.55
07/24/00	£ 772.92
07/31/00	£ 826.50

Starter	Cost	Number Sold	Income
Lentil Soup	£1.40	12	£16.80
Mulligatawny Soup	£1.50	4	£ 6.00
Prawn Cocktail	£2.10	9	£18.90
Shish Kebab	£3.95	12	£47.40
Sheek Kebab	£2.25	12	£27.00
Kebab	£2.25	10	£22.50
Lamb Tika	£2.25	13	£29.25
Chicken Tika	£2.25	25	£56.25
Tandoori Chicken	£2.25	39	£87.75
Meat Samosa	£1.80	18	£32.40
Vegetable Samosa	£1.60	10	£16.00
Onion Bhajee	£1.40	31	£43.40
Chicken Chatt	£2.25	11	£24.75
Chana Chatt	£2.10	3	£ 6.30
Prawn Puri	£2.35	10	£23.50
King Prawn Puri	£5.50	7	£38.50
Pakara	£1.90	12	£22.80
Stuffed Pepper	£2.95	9	£26.55
Tandoori Mixed Grill	£2.95	24	£70.80
Vegetable Korma	£2.10	23	£52.50
		294	£665.15 £781.55

7

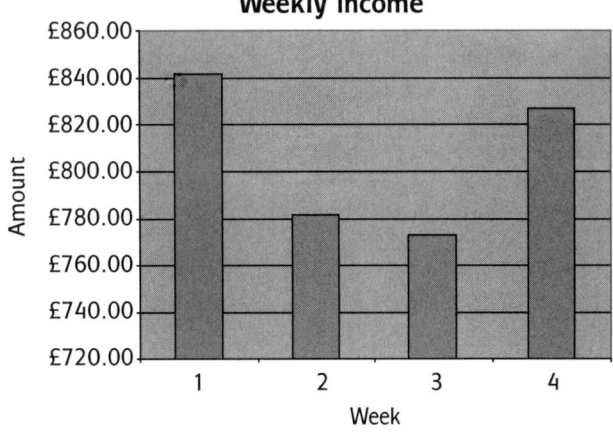

NA9: Analysing food

Requirements: internet, Excel
File: none
Aims:
■ use a given website to search for recipes
■ use the site for unit conversions, ratio and proportions.
NC Reference: KS4 Ma2 2f, 4a, 5f(F), 5g(H), (also Ma4, 1a)
Teacher preparation: check that there is access to the internet; website addresses.
Assumed knowledge:
■ copy data from the internet and paste it into a word processing package
■ edit data to reflect different proportion changes to ingredients
■ create bar [column] charts in Excel. Use simple formulae in Excel.
Support: EX2, EX3
Starting the activity: discuss how ratios are used in cooking.
Hints: to highlight columns that are not next to each other, hold down the Ctrl key on the keyboard.
Answers:

1 160g ≡ 5.643 oz; 75g ≡ 2.645 oz;
850g ≡ 29.982 oz; 150 ml ≡ 5.279 fl oz.

3

Fat in g per 100g
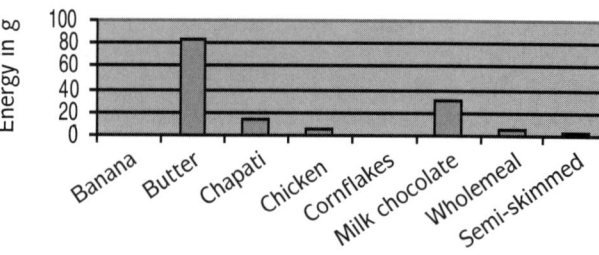

Protein in g per 100g
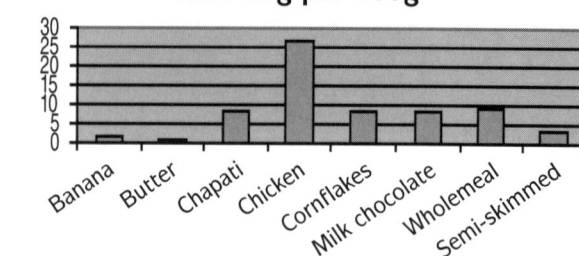

Calcium in mg per 100g
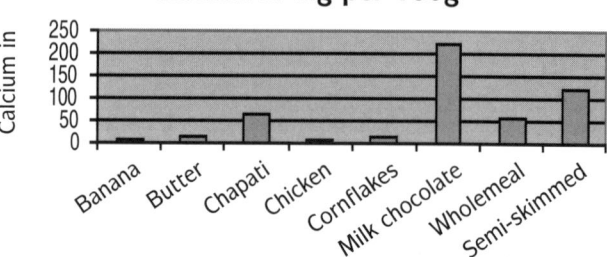

Energy produced in kJ per 100g
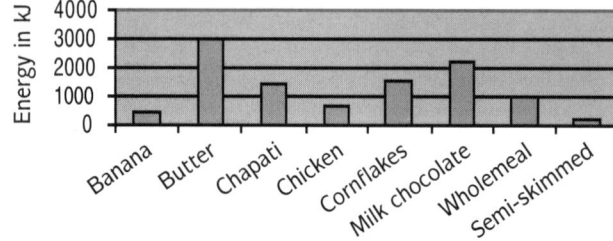

Carbohydrate in mg per 100g
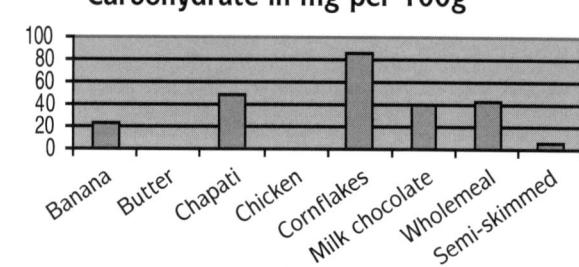

4

	Quantity	Energy produced in kJ	Fat in g	Protein in g	Calcium in mg	Carbohydrate in mg
Banana	2	806	0.6	2.4	12	46.4
Butter	1	30.31	0.817	0.005	0.15	0
Chapati	1	1383	12.8	8.1	66	48.3
Chicken	2	1198	8	53	18	0
Cornflakes	1.5	2302.5	1.05	11.85	22.5	128.85
Milk Chocolate	0	0	0	0	0	0
Wholemeal Bread	1	914	2.5	9.2	54	41.6
Semi-Skimmed milk	3	585	4.8	9.9	360	15
Totals		7218.81	30.567	34.455	532.65	280.15

NA10: Investigating quadratics 1

Requirements: Omnigraph
File: none
Aims:
■ use a graphing package to show how the values of the integers a, b and c affect the shape of the curve $y = ax^2 + bx + c$.
NC Reference: KS4 Ma2 5c(F), 5k(H), 6e(H) (also Ma1)
Teacher preparation: for Q4, have some prepared values for a, b and c.
Assumed knowledge:
■ Enter quadratic equations into Omnigraph.
Support: SSM14
Starting the activity:
Hints: Check that pupils are choosing their own values for a, b and c. If not suggest suitable values.
Answers:

1 **(a)** The curve has a minimum point.
  (b) The curve has a maximum point.

2 **(a)** The curve cuts above the x-axis.
  (b) The curve cuts below the x-axis.

3 **(a)** The curve cuts in two distinct places.
  (b) The curves touches the x-axis.
  (c) The curve does not cross the x-axis

4 **(a)** $a > 0$, $c > 0$, $b^2 - 4ac > 0$
  (b) $a > 0$, $c > 0$, $b^2 - 4ac < 0$
  (c) $a < 0$, $c > 0$, $b^2 - 4ac > 0$
  (d) $a > 0$, $c > 0$, $b^2 - 4ac = 0$
  (e) $a > 0$, $c > 0$, $b^2 - 4ac = 0$
  (f) $a > 0$, $c > 0$, $b^2 - 4ac = 0$

5 **(a)**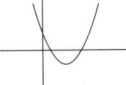

5 **(b)** Not possible to draw

5 **(c)**

5 **(d)**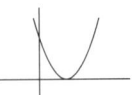

5 **(e)** Not possible to draw

5 **(f)**

NA11: Investigating sequences

Requirements: Excel, internet
File: none
Aims:
■ use a spreadsheet to generate the Fibonacci sequence and the Golden Number
■ look at a given website for information related to Fibonacci numbers.
NC Reference: KS4 Ma2 6a
Teacher preparation: check website addresses.
Assumed knowledge:
■ enter data and complicated formulae into Excel
■ produce line graphs in Excel.
Support: EX2
Starting the activity: discuss how the Golden ratio is used in architecture. Discuss convergence.
Hints: formulae can be copied by clicking on the black square at the bottom right of the cell and dragging the cell down to the required row. Remind pupils to check print preview before any printing.
Answers:

1 and **2**

The Fibonacci Sequence	The Ratio of Successive Terms
1	1.00000000000
1	2.00000000000
2	1.50000000000
3	1.66666666667
5	1.60000000000
8	1.62500000000
13	1.61538461538
21	1.61904761905
34	1.61764705882
55	1.61818181818
89	1.61797752809
144	1.61805555556
233	1.61802575107
377	1.61803713528
610	1.61803278689
987	1.61803444782
1597	1.61803381340
2584	1.61803405573
4181	1.61803396317
6765	1.61803399852
10946	1.61803398502
17711	1.61803399018
28657	1.61803398821
46368	1.61803398896
75025	

3

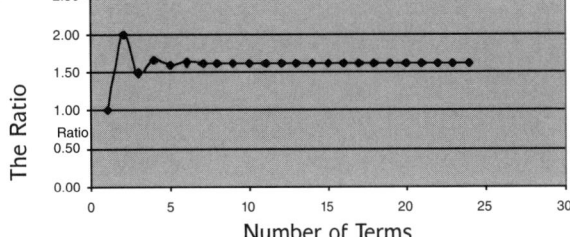

The Ratio of Successive Terms

5	The Fibonacci Sequence	The Ratio of Successive Terms		
	1	1.00000000000		
	1	2.00000000000		
	2	1.50000000000		
	3	1.66666666667		
	5	1.60000000000		
	8	1.62500000000		
	13	1.61538461538		
	21	1.61904761905		
	34	1.61764705882		
	55	1.61818181818		
	89	1.61797752809		
	144	1.61805555556		
	233	1.61802575107		
	377	1.61803713528		
	610	1.61803278689		
	987	1.61803444782		
	1597	1.61803381340		
	2584	1.61803405573		
	4181	1.61803396317		
	6765	1.61803399852		
	10946	1.61803398502		
	17711	1.61803399018		
	28657	1.61803398821	Sum of Highlight	
	46368	1.61803398896	3495921	
	75025	1.61803398867		
	121393	1.61803398878	Divide by 317811	
	196418	1.61803398874	11	
	317811	1.61803398875	The answer is	
	514229	1.61803398875	always 11	
	832040	1.61803398875		
	1346269	1.61803398875		
	2178309	1.61803398875		
	3524578	1.61803398875		
	5702887	1.61803398875		
	9227465	1.61803398875		
	14930352	1.61803398875		
	24157817	1.61803398875		
	39088169	1.61803398875		
	63245986	1.61803398875		
	102334155	1.61803398875		
	165580141	1.61803398875		
	267914296	1.61803398875		
	433494437	1.61803398875		
	701408733	1.61803398875		
	1134903170	1.61803398875		
	1836311903	1.61803398875		
	2971215073	1.61803398875		
	4807526976	1.61803398875		
	7778742049			

NA12: Linear functions and their inverse

Requirements: Omnigraph
File: none
Aims:
- use a graphing package to help pupils find the inverse of linear functions.

NC Reference: KS4 Ma2 6b, 6c, 6d(H), 6e(F)
Teacher preparation:
Assumed knowledge:
- Enter linear equations, some containing fractions, into Omnigraph.

Support: SSM14
Starting the activity: discuss reflections of lines.
Hints: use the fraction icon to enter 1/2.
Answers:

1 (a) $y = \frac{x}{2}$

(b) $y = \frac{x+1}{2}$

(c) $y = \frac{x-4}{2}$

(d) $y = 2(x-1)$

(e) $y = \frac{x+1}{3}$

(f) $y = \frac{x-2}{3}$

(g) $y = \frac{-x}{2}$

(h) $y = \frac{x-3}{-2} \equiv \frac{3-x}{-2}$

(i) $y = 3 - x$

(j) $y = 4 - x$

2 These are self-inverse functions because they produce the same graph. They are reflecting each other in the line $y = x$.

3 Any correct two, for example, $y = 6 - x$ and $y = \frac{3}{x}$.

4 $y = \frac{x-c}{m}$

NA13: Investigating quadratics 2

Requirements: Omnigraph
File: none
Aims:
- use a graphing package to produce quadratic curves by completing the square and using translations.

NC Reference: KS4 Ma2 5k(H), 6e
Teacher preparation:
Assumed knowledge:
- enter quadratic equations into Omnigraph.

Support: SSM14
Starting the activity: remind pupils about translations.
Hints: the screen needs to be cleared before starting each part of the question. Answers can be checked on screen. To enter the cubed (or squared) sign in Omnigraph, hold down the Alt key on the keyboard as you press 3 (or 2). Alternatively you can type x^3 (or x^2) or use the 'power' icon.

Answers:

1 **(a)** The curve has a turning point at $(1, 0)$.
   (b) The curve has been translated 1 unit to the right.

2 **(a)** The curve has a turning point at $(-2, 0)$.
   (b) The curve has been translated 2 units to the left.

3 **(a)** The curve has a turning point at $(-5, 0)$.
   (b) The curve has a turning point at $(3, 0)$.
   (c) The curve has a turning point at $(4, 0)$.
   (d) The curve has a turning point at $(2, 0)$.
   (e) The curve has a turning point at $(-3, 0)$.

4 **(a)** $y = (x + 4)^2$
   (b) $y = (x - 5)^2$
   (c) $y = (x + 3)^2$
   (d) $y = (x - 1.5)^2$
   (e) $y = (x + 2.5)^2$

5 The curve has a turning point at $(3, 2)$.
   The transformations are a translation 3 units to the right and another 2 places up.

6 **(a)** The minimum point is at $(-3, -2)$.
   (b) $y = x^2 + 6x + 7$

7 **(a)** $y = (x - 4)^2 + 2 = x^2 - 8x + 18$
   (b) $y = (x + 5)^2 + 2 = x^2 + 10x + 27$
   (c) $y = (x - 4)^2 - 2 = x^2 - 8x + 14$
   (d) $y = (x + 4)^2 - 3 = x^2 + 8x + 13$
   (e) $y = (x + 6)^2 + 5 = x^2 + 12x + 41$

8 **(a)** $y = (x - 1)^2 - 3$
   (b) $y = (x + 1)^2 + 6$
   (c) $y = (x + 3)^2 - 5$
   (d) $y = (x - 2)^2 - 6$
   (e) $y = (x - 5)^2 - 46$

NA14: Approximating a value for π

Requirements: Excel, internet
File: none
Aims:
- use a spreadsheet and the internet to investigate infinite series, sequences, convergence, percentage errors and other types of numbers
- research given sites for information related to π and irrational, transcendental and imaginary numbers.

NC Reference: KS4 Ma2 1d, 2a, 2b, 2c, 2d, 6a(H) (also Ma1)
Teacher preparation: check that there is access to the internet; website addresses.
Assumed knowledge:
- enter formulae into Excel.

Support: EX2
Starting the activity: discuss why π is irrational.
Hints: it is a good idea to plan each spreadsheet on paper first. If a pupil wishes to print then suggest they only print one page of data.

Answers:

1 After 2000 terms an estimate is 3.140885547.

Numerator	Denominator	Decimal Fraction	Cumulative Sum
1	1	2.828427125	2.828427125
1	3	0.942809042	3.771236166
-1	5	-0.565685425	3.205550741
1	7	-0.404061018	2.801489724
-1	9	0.314269681	3.115759404
1	11	0.257129739	3.372889143

2 After 2000 terms an estimate is 3.14080775.

Numerator	Denominator	Decimal Fraction	Cumulative Product
2	1	4.000000000	4.000000000
2	3	0.666666667	2.666666667
4	3	1.333333333	3.555555556
4	5	0.800000000	2.844444444
6	5	1.200000000	3.413333333
6	7	0.857142857	2.925714286
8	7	1.142857143	3.343673469
8	9	0.888888889	2.972154195
10	9	1.111111111	3.302393550
10	11	0.909090909	3.002175955

3 After 2000 terms an estimate is 3.141115272.

4

$$Percentage\ error = \frac{|Estimate - Original|}{Original} \times 100$$

$$Percentage\ error = \frac{|3.140885547 - 3.141592654|}{3.141592654} \times 100 = 0.02\%$$

$$Percentage\ error = \frac{|3.140807746 - 3.141592654|}{3.141592654} \times 100 = 0.02\%$$

$$Percentage\ error = \frac{|3.141115272 - 3.141592654|}{3.141592654} \times 100 = 0.02\%$$

5 After 2000 terms the sum of the series is 1.9990005. After 2000 terms the sum of the series is 1.999900005. The series is slowly converging to 2.

Term	Denominator	Decimal Fraction	Cumulative Sum
1	1	1	1
2	3	0.333333333	1.33333333
3	6	0.166666667	1.5
4	10	0.1	1.6
5	15	0.066666667	1.66666667
6	21	0.047619048	1.71428571

NA15: Infinite series

Requirements: Excel
File: none
Aims:
■ use a spreadsheet to investigate the divergence and convergence of infinite series.
NC Reference: KS4 Ma2 6a(H) (also Ma1)
Teacher preparation:
Assumed knowledge:
■ enter formulae into Excel.
Support: EX2
Starting the activity: discuss the difference between convergence and divergence. Start the harmonic series on a calculator before launching Excel.
Hints: plan each spreadsheet on paper first.
Answers:

3	0.333333333	1.833333333
4	0.25	2.083333333
5	0.2	2.283333333
6	0.166666667	2.45
7	0.142857143	2.592857143
8	0.125	2.717857143
9	0.111111111	2.828968254
10	0.1	2.928968254
11	0.090909091	3.019877345
12	0.083333333	3.103210678
13	0.076923077	3.180133755
14	0.071428571	3.251562327
15	0.066666667	3.318228993
16	0.0625	3.380728993
17	0.058823529	3.439552523
18	0.055555556	3.495108078
19	0.052631579	3.547739657
20	0.05	3.597739657

1 The approximate sum of the first 20 terms is 3.59773965714368.

2 The approximate sum of the first 200 terms is 5.87803094812145.

3

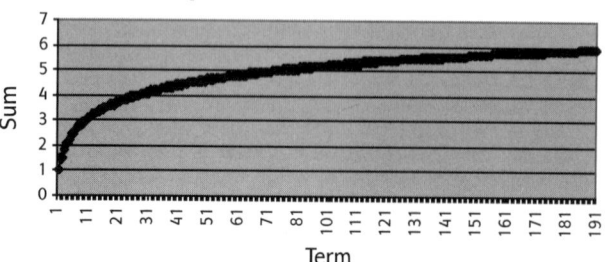

Line Graph of The Cumulative Sum

The series is diverging very slowly.

4 The approximate sum of the first 2000 terms is 8.178368104361028.

5 The sum passes 10 after 12367 terms.

6 The sum passes 11 after 33616 terms.

7

Denominator	Numerator	Decimal Fraction	Cumulative Sum
1	1	1	1
2	−1	−0.5	0.5
3	1	0.333333	0.833333
4	−1	−0.25	0.583333
5	1	0.2	0.783333
6	−1	−0.16667	0.616667
7	1	0.142857	0.759524
8	−1	−0.125	0.634524
9	1	0.111111	0.745635
10	−1	−0.1	0.645635
11	1	0.090909	0.736544
12	−1	−0.08333	0.653211
13	1	0.076923	0.730134
14	−1	−0.07143	0.658705
15	1	0.066667	0.725372
16	−1	−0.0625	0.662872
17	1	0.058824	0.721695
18	−1	−0.05556	0.66614
19	1	0.052632	0.718771
20	−1	−0.05	0.668771

8

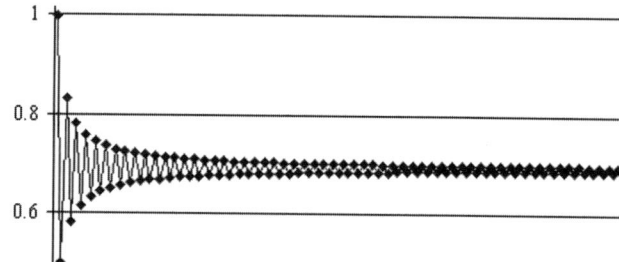

Line Graph of The Cumulative Sum

9 This series converges to 2.

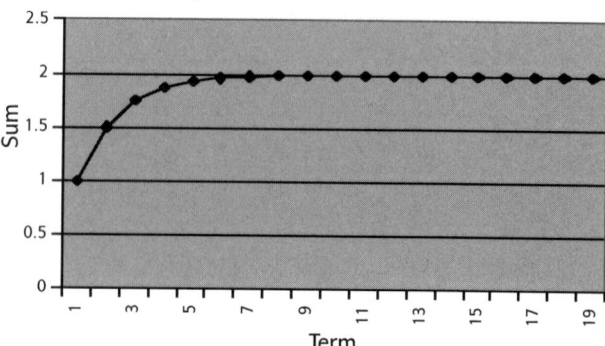

Line Graph of The Cumulative Sum

10 (a) The series converges.
(b) The series diverges.
(c) The series diverges.
(d) This is the diverging series investigated in questions **1** – **6**

NA16: Solving equations

Requirements: Omnigraph
File: none
Aims:
■ use a graphing package to solve quadratic and cubic equations.
NC Reference: KS4 Ma2 5k(H), 5l(H), 6f(H) but ext. to cubics
Teacher preparation:
Assumed knowledge:
■ enter cubic equations into Omnigraph
Support: NA13
Starting the activity: discuss the general shape of quadratics and cubics.
Hints: to enter the cubed (or squared) sign in Omnigraph, hold down the Alt key on the keyboard as you press 3 (or 2). Alternatively you can type x^3 (or x^2) or use the 'power' icon.
Answers:

1 (a) Solutions to the equation are $x = -1$, $x = 2$ and $x = 3$.

    (b) Solutions to the equation are approximately $x = -0.8$, $x = 1.5$ and $x = 3.38$.

2 (a) Solutions to the equation are $x = -2$, $x = -1$ and $x = \frac{1}{2}$.

    (b) Solution to the equation is approximately $x = -2.4$.

3 (a) Solutions to the equation are $x = 1$ and $x = 4$. There are only two as one root is repeated at $x = 1$.

    (b) and (c) Solution to the equation is approximately $x = 4.5$.

4 (a) Solutions to the equation are approximately $x = -2$, $x = -1$ and $x = 1$.

    (b) Solutions to the equation are approximately $x = -2.6$, $x = -0.4$ and $x = 1$.

    (c) Solutions to the equation are approximately $x = -2.6$, $x = -0.4$ and $x = 1$.

    (d) Solutions to the equation are approximately $x = -2.7$, $x = -0.8$ and $x = 1.5$.

NA17: The approximate solution to equations

Requirements: Excel
File: none
Aims:
■ use a spreadsheet to solve equations by iteration.
NC Reference: KS4 Ma2 5f(F), 5g(H), 5m(H)
Teacher preparation: work through the activity sheet example.
Assumed knowledge:
■ enter complicated formulae in Excel.
Support: NA11
Starting the activity: discuss iterative methods.
Hints: To key in x^2 type x then hold down the Alt key on your keyboard as you type 253 on the number pad. Make sure that the Num Lock is on.

Explain the importance of brackets in the formulae.
Answers:

x	$(5 - x^2)/7$
0.5	0.678571429
0.678571429	0.648505831
0.648505831	0.654205741
0.654205741	0.653144978
0.653144978	0.653343091
0.653343091	0.653306115
0.653306115	0.653313017
0.653313017	0.653311729
0.653311729	0.653311969
0.653311969	0.653311924
0.653311924	0.653311933
0.653311933	0.653311931
0.653311931	0.653311932
0.653311932	0.653311931
0.653311931	0.653311931

1

x	$(x^2 + 6)/9$
1	0.777777778
0.777777778	0.73388203
0.73388203	0.726509204
0.726509204	0.725312847
0.725312847	0.725119858
0.725119858	0.725088757
0.725088757	0.725083745
0.725083745	0.725082937
0.725082937	0.725082807
0.725082807	0.725082786
0.725082786	0.725082783
0.725082783	0.725082782
0.725082782	0.725082782
0.725082782	0.725082782
0.725082782	0.725082782

2

x	SQRT$(7x - 2)$
7	6.8556546
6.8556546	6.781561929
6.781561929	6.743213886
6.743213886	6.723280241
6.723280241	6.712895179
6.712895179	6.707478383
6.707478383	6.704651272
6.704651272	6.703175285
6.703175285	6.702404568
6.702404568	6.702002087
6.702002087	6.701791896
6.701791896	6.701682122
6.701682122	6.701624792
6.701624792	6.701594851
6.701594851	6.701579214

3

x	$(x\hat{\ }3 + 2x\hat{\ }2 -1)/5$
0	0.2
0.2	0.1824
0.1824	0.185478415
0.185478415	0.184962928
0.184962928	0.185049922
0.185049922	0.18503526
0.18503526	0.185037732
0.185037732	0.185037315
0.185037315	0.185037385
0.185037385	0.185037374
0.185037374	0.185037376
0.185037376	0.185037375
0.185037375	0.185037375
0.185037375	0.185037375
0.185037375	0.185037375

4

x	$3\text{root}(15 -3x)$
2	2.080083283
2.080083283	2.061407632
2.061407632	2.065593312
2.065593312	2.064765108
2.064765108	2.065006258
2.065006258	2.064949705
2.064949705	2.064962967
2.064962967	2.064959857
2.064959857	2.064960586
2.064960586	2.064960415
2.064960415	2.064960455
2.064960455	2.064960446
2.064960446	2.064960448
2.064960448	2.064960448
2.064960448	2.064960448

5

x	$4\text{root}(4x + 3)$
1.7	1.769320539
1.769320539	1.781705229
1.781705229	1.783890869
1.783890869	1.784275756
1.784275756	1.784343508
1.784343508	1.784355434
1.784355434	1.784357533
1.784357533	1.784357902
1.784357902	1.784357967
1.784357967	1.784357979
1.784357979	1.784357981
1.784357981	1.784357981
1.784357981	1.784357981
1.784357981	1.784357981
1.784357981	1.784357981

NA18: Odd and even functions

Requirements: Omnigraph
File: none
Aims:
■ use a graphing package to investigate odd and even functions.
NC Reference: KS4 Ma2 1j, 6f(H) (also Ma1)
Teacher preparation: familiarise yourself with switching between polar trigonometric or default axes in Omnigraph.
Assumed knowledge:
■ enter a variety of equations into Omnigraph
■ use trigonometric axes.
Support: NA13
Starting the activity: discuss reflections and rotations of curves.
Hints: functions will need to be cleared between questions.
When entering trigonometric functions, change the axes to trigonometric by clicking <u>Z</u>oom and then <u>T</u>rig scales.
To change back click <u>Z</u>oom and then <u>R</u>eset scales.
Answers:

1 (a) $y = x^n$ is an odd function for odd values of n
   (b) $y = x^n$ is even for even values of n.

2 (a) The product of two even functions is even.
   (b) The difference of two even functions is even.
   (c) The product of two odd functions is even.
   (d) The product of an odd function and an even function is odd.

3

Function	Odd or Even
$y = \sin x$	Odd
$y = \cos x$	Even
$y = \sin x \cos x$	Odd
$y = \dfrac{1}{x^2}$	Even
$y = \dfrac{1}{x}$	Odd
$y = x - \dfrac{1}{x}$	Odd
$y = x^2 + \dfrac{1}{x^2}$	Even
$y = x^3 + \dfrac{1}{x^3}$	Odd
$y = x^2 + \dfrac{1}{x^3}$	Neither odd nor even

NA19: Function transformations

Requirements: Omnigraph
File: none
Aims:
■ use a graphing package to investigate function transformations.
NC Reference: KS4 Ma2 6f(H), 6g(H)
Assumed knowledge:
■ enter a variety of equations into Omnigraph
■ use trigonometric axes.
Support: NA13
Starting the activity: discuss the transformations used in the activity.
Hints: remind the pupils of the buttons they need to click to clear functions, add a fraction and add powers.
Answers:

1

New curve	Transformations
$y \sin 2x$	A stretch parallel to the x-axis; scale factor of $\frac{1}{2}$.
$y = 2\sin x$	A stretch parallel to the y-axis; scale factor of 2.
$y = \sin\frac{1}{2}x$	A stretch parallel to the x-axis; scale factor of 2.
$y = \frac{1}{2}\sin x$	A stretch parallel to the y-axis; scale factor of $\frac{1}{2}$.
$y = \sin x + 3$	A translation parallel to the y-axis; 3 units up.
$y = \sin x - 4$	A translation parallel to the y-axis; 4 units down.
$y = 3\sin\frac{1}{2}x$	A stretch parallel to the x-axis, scale factor of 2 and a stretch parallel to the y-axis, scale factor of 3.
$y = \frac{1}{3}\sin 3x - 2$	A stretch parallel to the x-axis, scale factor of $\frac{1}{3}$ and a stretch parallel to the y-axis, scale factor of $\frac{1}{3}$ followed by a translation parallel to the y-axis, 2 units down.

2

New Curve	Transformation(s) from $y=2^x$	Transformation(s) from $y=x^3$
$y = 2^{-x}$	Reflection in the y-axis	
$y = -2^x$	Reflection in the x-axis	
$y = (x-3)^3$		Translation parallel to the x-axis, 3 units to the right.
$y = (x+2)^3$		Translation parallel to the x-axis, 2 units to the left.
$y = 2^x + 1$	Translation parallel to the y-axis, 1 unit up.	
$y = (x-2)^3 + 1$		Translation parallel to the x-axis, 2 units to the right and a translation parallel to the y-axis, 1 unit up.
$y = (x+2)^3 - 1$		Translation parallel to the y-axis 2 units to the left and a translation parallel to the x-axis 1 unit down.
$y = 2^{2x}$	A stretch parallel to the x-axis; scale factor of $\frac{1}{2}$	

3 **(a)** Translation parallel to the x-axis a units to the left.
(b) Translation parallel to the x-axis b units to the right.
(c) Translation parallel to the y-axis a units up.
(d) Translation parallel to the y-axis b units down.
(e) Stretch parallel to the y-axis by a factor of a.
(f) Stretch parallel to the x-axis by a factor of $1/a$.

4 **(a)** Reflection in the x-axis.
(b) Reflection in the y-axis.

SSM1: Plotting coordinates 1

Requirements: Omnigraph
File: none
Aims:
■ use a graphing package to draw polygons.
NC Reference: KS3 Ma3 1a, 2g, 3e
Teacher preparation: familiarise yourself with how Omnigraph is used to create shapes.
Assumed knowledge:
■ entering coordinates in Omnigraph.

Support:
Starting the activity: discuss the various polygons used.
Hints: sketch shapes on paper first.
Answers: none

SSM2: Plotting coordinates 2

Requirements: Omnigraph, access to a printer
File: none
Aims:
■ use a graphing package to create polygons and lines of symmetry.
NC Reference: KS3 Ma3 1a, 1d, 1e, 2g, 3e
Teacher preparation: check that pupils have access to a printer. Have some additional shapes for extension work.
Assumed knowledge:
■ create polygons in Omnigraph.
Support: SSM1
Starting the activity: discuss lines of symmetry.
Hints: to draw lines in Omnigraph, select <u>S</u>hapes and then the appropriate polygon.
Use the hand button to move the grid as required. Horizontal and vertical lines can be drawn by selecting <u>A</u>nalysis and then <u>S</u>traight lines.
Answers:

3 and **4**

(a) concave quadrilateral

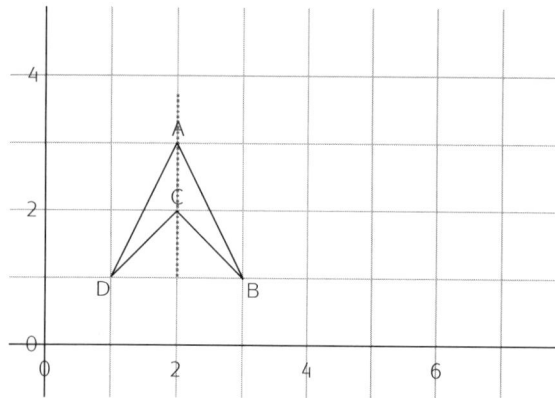

(b) convex pentagon

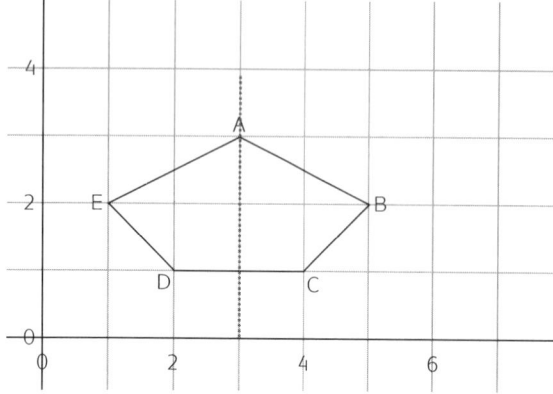

(c) parallelogram

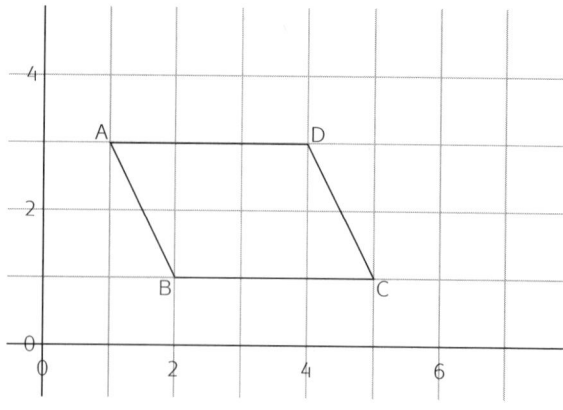

(d) concave hexagon

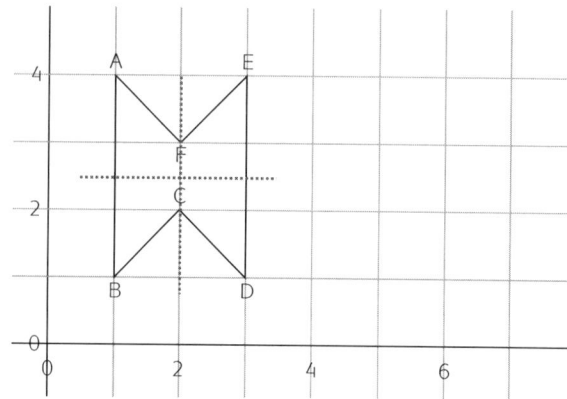

(e) concave octogon

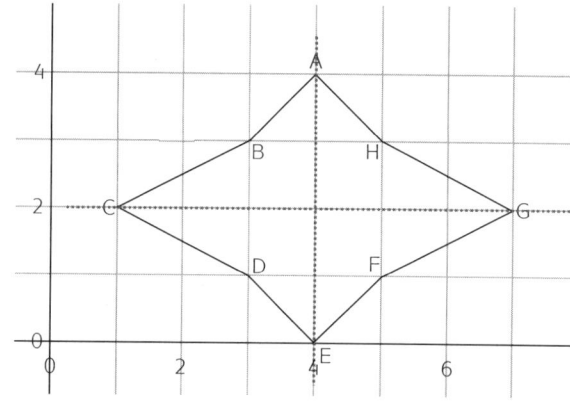

SSM3: Investigating angles 1

Requirements: The Geometer's Sketchpad
File: none
Aims:
■ use a dynamic geometry package to investigate angles on a straight line and angles round a point.
NC Reference: KS3 Ma3 2a
Teacher preparation: familiarise yourself with The Geometer's Sketchpad and check the default settings using <u>D</u>isplay, <u>P</u>references. Adjust as necessary.
Assumed knowledge:
■ none.
Support:
Starting the activity: explain the requirements of an investigation.

Hints: the labels on screen may be different to those on the activity sheet. To measure angles, the arrow in the menu bar should be selected. The points need to be selected in the correct order. Points may need to be deselected first (see p 22).

Answers:

1 The sum of the angles ACD and BCD is always 180°.

2 Yes the sum remains the same.

3 Angles FJH and IJG are the same.

4 Angles FJI and GJH are the same.

5 Yes the angles remain the same.

6 The sum of ACD, DCF, FCB, BCE and ECD is 360°.

7 The sum of the angles remains 360°.

SSM4: Estimating angles

Requirements: The Geometer's Sketchpad
File: none
Aims:
■ use a dynamic geometry package to estimate and check the size of acute, obtuse and reflex angles.
NC Reference: KS3 Ma3 2b
Teacher preparation: familiarise yourself with The Geometer's Sketchpad and check the default settings using <u>D</u>isplay, <u>P</u>references. Adjust as necessary.
Assumed knowledge:
■ draw lines and measure angles in The Geometer's Sketchpad.
Support: SSM3
Starting the activity: discuss the different types of angles.
Hints: the calculator can be used to display the subtraction.
Answers: none

SSM5: Drawing angles and lines

Requirements: MSWLogo
File: none
Aims:
■ use Logo to draw acute, obtuse and reflex angles, parallel and perpendicular lines, equilateral triangles, squares and parallelograms.
NC Reference: KS3 Ma3 2b, 2c, 2f, 4j
Teacher preparation: prepare some extension material using Logo to create some images, e.g.

Assumed knowledge:
■ use simple MSWLogo commands to create acute, obtuse and reflex angles, parallel and

perpendicular lines and some related polygons.
Support:
Starting the activity:
Hints: It is a good idea to sketch the angles on paper first. It is suggested that the arms of all the drawings are 100 units long.
In Q4, make use of penup or pu to move between letters and pendown or pd before drawing your next letter.
Answers: none

SSM6: Investigating angles 2

Requirements: The Geometer's Sketchpad
File: none
Aims:
■ use a dynamic geometry package to investigate corresponding and alternate angles and parallelograms.
NC Reference: KS3 Ma3 2c, 2f, 2g
Teacher preparation: familiarise yourself with The Geometer's Sketchpad and check the default settings using <u>D</u>isplay, <u>P</u>references. Adjust as necessary.
Assumed knowledge:
■ draw lines and measure angles in The Geometer's Sketchpad.
Support: SSM3
Starting the activity: discuss the various types of angles used.
Hints: to measure angles, the arrow in the menu bar should be selected. The points need to be selected in the correct order. Points may need to be deselected first (see p 22).
Answers:

1 They are equal.

2 Yes, the three angles remain equal.

3 They are equal.

4 Yes, they remain equal.

5 Their sum is 180°, they are supplementary.

6 The sum is 180°.

7 The two pairs of supplementary angles BAC, ACD and ABD, BDC total 360°.

8 Yes, the sum remains the same.

SSM7: Investigating angles and triangles in quadrilaterals

Requirements: The Geometer's Sketchpad
File: none
Aims:
■ use a dynamic geometry package to investigate the angles in triangles and quadrilaterals.
NC Reference: KS3 Ma3 2c, 2d, 2f, 2g
Teacher preparation: familiarise yourself with The Geometer's Sketchpad and check the default settings using <u>D</u>isplay, <u>P</u>references. Adjust as necessary.

Assumed knowledge:
- create lines, triangles and quadrilaterals in The Geometer's Sketchpad.

Support: SSM6

Starting the activity: explain the requirements of an investigation.

Hints: When dragging a vertex, all angles in the quadrilateral should be kept to or less than 180° as The Geometer's Sketchpad does not measure reflex angles. To measure angles, the arrow in the menu bar should be selected. The points need to be selected in the correct order. Points may need to be deselected first (see p 22).

Answers:

1 Sum of the three angles is 180°.

2 Yes, the sum remains 180°.

3 The sum remains the same.

4 The sum of angles BAC and ACB is the same as angle DBC.

5 and **6** The same result as part 4 applies.

7 The sum of the angles is 360° because the quadrilateral is made from the two triangles ABC and ACD.

SSM8: Imagine

Requirements: MSWLogo, internet

File: infinity.lgo

Aims:
- use Logo and the internet to investigate infinity by looking at area and perimeter.

NC Reference: KS3 Ma3 2e, 4f, 4j (also Ma1)

Teacher preparation: check that pupils can access the file; that there is access to the internet; website addresses; alter the procedure to create 'inverted snowflakes'.

Assumed knowledge:
- enter commands into a given MSWLogo file.

Support: SSM5

Starting the activity: discuss how a shape can have a finite area but infinite perimeter.

Hints: anything beyond snow 8 is beyond the precision of the software.

Answers: none

SSM9: Coordinates, translations, reflections and rotations

Requirements: Omnigraph

File: none

Aims:
- use a graphing package to look at translations, reflections and rotations.

NC Reference: KS3 Ma3 3a, 3b, 3e

Teacher preparation:

Assumed knowledge:
- enter coordinates to create shapes and apply transformations in Omnigraph.

Support: SSM2

Starting the activity: discuss the various transformations or vectors.

Hints: –90° is clockwise and 90° is anticlockwise in Omnigraph rotations.

Answers:

1 (a) B = (1,3), C = (2,2) and D = (1,2)

   (b) A′ = (3,0), B′ = (3,2), C′ = (4,1) and D′ = (3,1)

   (c) Enter the vector $\left(\begin{smallmatrix} 2 \\ -1 \end{smallmatrix}\right)$

   (d) A′ = (–2,3), B′ = (–2,5), C′ = (–1,4) and D′ = (–3,4)

   (e) The required vector is $\left(\begin{smallmatrix} -3 \\ 5 \end{smallmatrix}\right)$

   (g) (i) The required vector is $\left(\begin{smallmatrix} 3 \\ -2 \end{smallmatrix}\right)$

   (g) (ii) The required vector is $\left(\begin{smallmatrix} 0 \\ 4 \end{smallmatrix}\right)$

2 (a) A′ = (–1,1), B′ = (–1,3), C′ = (–2,2) and D′ = (–1,2)

   (c) A′ = (1,–1), B′ = (1,–3), C′ = (2,–2) and D′ = (1,–2)

   (d) (i) A′ = (3,1), B′ = (3,3), C′ = (2,2) and D′ = (3,2)

   (d) (ii) A′ = (1,1), B′ = (1,–1), C′ = (2,0) and D′ = (1,0)

   (d) (iii) A′ = (1,1), B′ = (3,1), C′ = (2,2) and D′ = (2,1)

3 (a) (i) A′ = (1,1), B′ = (3,1), C′ = (2,0) and D′ = (2,1)

   (a) (ii) A′ = (1,1), B′ = (–1,1), C′ = (0,2) and D′ = (0,1)

   (a) (iii) A′ = (1,1), B′ = (1,–1), C′ = (0,0) and D′ = (1,0)

   (a) (iv) A′ = (1,1), B′ = (1,–1), C′ = (0,0) and D′ = (1,0)

   (b) A rotation in either direction of 180° produces the same result.

SSM10: Coordinates, enlargements and ratios

Requirements: Omnigraph

File: none

Aims:
- use a graphing package to look at enlargements and ratios.

NC Reference: KS3 Ma3 3a, 3c, 3d, 3e

Teacher preparation:

Assumed knowledge:
- enter coordinates to create shapes and apply enlargements in Omnigraph.

Support: SSM2

Starting the activity: discuss enlargements (positive and negative) and ratios.

Hints:

Answers:

1 (a) A′ = (2,2), B′ = (0,2), C′ = (0,0) and D′ = (2,0)

   (b) A′ = (4,4), B′ = (0,4), C′ = (0,0) and D′ = (4,0)

(d) $A' = (6,6)$, $B' = (0,6)$, $C' = (0,0)$ and $D' = (6,0)$

(e) $A' = (3,3)$, $B' = (–1,3)$, $C' = (–1,–1)$ and $D' = (3,–1)$

(f) $A' = (2,4)$, $B' = (–2,4)$, $C' = (–2,0)$ and $D' = (2,0)$

(g) $A' = (3,2)$, $B' = (-1,2)$, $C' = (–1,–2)$ and $D' = (3,-2)$

(h) $A' = (–1,2)$, $B' = (3,2)$, $C' = (3,6)$ and $D' = (-1,6)$

(i) $A' = (–6,–6)$, $B' = (0,–6)$, $C' = (0,0)$ and $D' = (–6,0)$

2 (a)

Scale factor	Ratio of lengths	Ratio of areas
2	2	4
3	3	9
4	4	16
–2	2	4
–3	3	9
0.5	0.5	0.25
–0.5	0.5	0.25

(b) The ratio of lengths changes in proportion to the enlargement factor. The ratio of areas changes in proportion to the square of the enlargement factor.

SSM11: Coordinates and straight lines

Requirements: Omnigraph
File: none
Aims:
- use a graphing package to look at the equations of horizontal and vertical lines.

NC Reference: KS3 Ma3 1a, 3e (also Ma2 6e)
Teacher preparation:
Assumed knowledge:
- enter coordinates in Omnigraph.

Support: SSM1
Starting the activity: ask pupils to describe the difference between horizontal and vertical lines.
Hints: To draw a line, select <u>S</u>hapes and then <u>N</u>ew shape.
May need to remind pupils of which button to press to start a new question.
Answers:

1 Coordinates include: (2,2), (2,3), (2,4), (2,5) and (2,6).

2 Coordinates include: (2,4), (3,4) and (4,4).

3 Coordinates include: (–1,2), (–1,1), (–1,0), (–1,–1), (–1,–2) and (–1,–4).

4 Coordinates include: (–1,–2), (0,–2), (1,–2), (2,–2) and (3,–2).

5 (e) (3,5), (3,–4), (–4,5), (–4,4)

6 (b) Coordinates include: (1,1), (2,2) and (3,3).

(d) Coordinates include: (1,–1), (–1,1) and (2,–2).

7 (a) The line $y = 0$.

(b) The line $x = 0$.

SSM12: Patterns

Requirements: Omnigraph, internet
File: none
Aims:
- use a graphing package to create unusual symmetrical shapes using common trigonometric functions
- introduces polar coordinates
- look at other curves at a given website.

NC Reference: KS3 Ma3 1d, 1f, 1g, 1i, 1j, 4j
Teacher preparation: check that there is access to the internet; website addresses.
Assumed knowledge:
- enter a variety of polar equations, mainly trigonometric and involving fractions, into Omnigraph.

Support: SSM2
Starting the activity: discuss the need for different coordinate systems.
Hints: in question 1 etc., you do not need to clear the screen to type new equations. Just highlight the multiple of θ, change the number and press the Enter key.
An alternative way to investigate, for example, $r = 1–\cos n\theta$ is to define the constant $a = 1$ and use the Dynamic Constant Editor, with step 1, to run through $r = 1–\cos a\theta$. Note Omnigraph does not accept n as a constant.
Answers:

1 Increasing the value of n increases the number of loops .

2 Increasing the value of n increases the number of loops.

4 Increasing the value of n increases the number of loops above and below the x-axis. The number n indicating the number of loops each side.

5 For even values of n we obtain two loops. For odd values of n we obtain only one loop.

6 When both n and m are equal to 2 a circle of radius 1 is produced. For even values a 'star' is produced. For odd values a 'kidney' is produced.

7

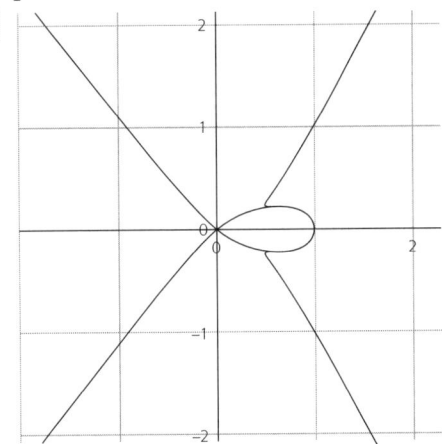

10

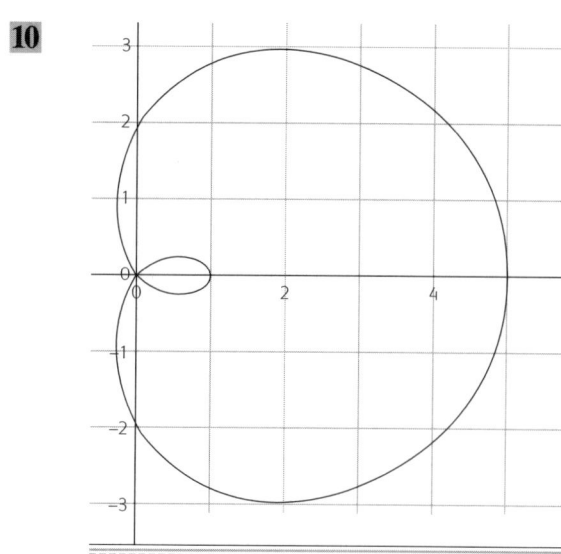

$r = 2 + 3\cos\theta$ for 1 revs

SSM13: Investigating straight lines 1

Requirements: The Geometer's Sketchpad
File: none
Aims:
■ use a dynamic geometry package to investigate the gradients of parallel and perpendicular lines to understand the equation $y = mx + c$.
NC Reference: KS3 Ma3 2a, 2c (also Ma2 6e, 6g, 6h)
Teacher preparation: check out the change of precision required on the activity sheet.
Assumed knowledge:
■ create lines, measure slope and add axes in The Geometer's Sketchpad.
Support: SSM3
Starting the activity: mention that 'gradient' is 'slope' in The Geometer's Sketchpad.
Hints:
Answers:

1 Positive gradient; line slopes up to right; negative gradient: slopes down to right.

2 Horizontal line.

3 Line gets steeper and closer to vertical.

4 Infinity.

5 The product of their gradients is –1.

6 They are the same.

SSM14: Investigating straight lines 2

Requirements: Omnigraph
File: none
Aims:
■ use a graphing package to produce straight lines by a series of transformations – change $y = x$ into $y = mx + c$.
NC Reference: KS3 Ma3 3a, 3b, 4j
Teacher preparation:
Assumed knowledge:
■ enter linear equations, sometimes involving fractions, into Omnigraph.
Support: SSM9

Starting the activity: review horizontal and vertical lines and introduce an equation like $y = 5x$.
Hints: you may need to remind pupils to clear the screen before each part of the question.
Pupils should use each line that they draw to help them work out how to draw the next line.
Answers:

1 a = 0, b = 1
2 a = 0, b = 2
3 a = 0, b = 3
4 **(a)** $y = 4x$
  (b) $y = 5x$
5 g = 0, h = 2
6 k = 0, l = 4
7

Equation of the line	Passing through
$y = 8x$	(0,0) and (1,8)
$y = 6x$	(0,0) and (1,6)
$y = \frac{1}{3}x$	(0,0) and (3,1)
$y = 10x$	(0,0) and (1,10)
$y = \frac{1}{5}x$	(0,0) and (5,1)

13 **(a)** Curve to enter $y = -5x+1$
  (b) Curve to enter $y = -3x+4$
  (c) Curve to enter $y = 4x+2$
  (d) Curve to enter $y = \frac{-1}{3}x+1$
  (e) Curve to enter $y = \frac{1}{2}x+1$

SSM15: Drawing bearings

Requirements: MSWLogo   *File*: none
Aims:
■ use Logo to draw the points of the compass and bearings.
NC Reference: KS3 Ma3 4b, 4j; KS4 Ma3 4a(H), 4b(F), 4e(H), 4j(F)
Teacher preparation:
Assumed knowledge:
■ use simple MSWLogo commands.
Support: SSM5
Starting the activity: discuss the difference between the way angles and bearings are represented.
Hints: there is no Undo command in MSWLogo. The best 'solution' is to change the pen colour to white and reverse the incorrect instruction.
Answers: none

SSM16 Investigating circles 1

Requirements: The Geometer's Sketchpad   *File*: none
Aims:
■ use a dynamic geometry package to investigate the tangents and chords of a circle.
NC Reference: KS4 Ma3 2h(H), 2i(F)
Teacher preparation: familiarise yourself with The Geometer's Sketchpad and check the default settings using Display, Preferences. Adjust as necessary.
Assumed knowledge:
■ draw lines and measure angles in The Geometer's Sketchpad.

Support: SSM13
Starting the activity: explain the requirements of an investigation.
Hints: the labels on screen may be different to those on the activity sheet. To measure angles, the arrow in the menu bar should be selected. The points need to be selected in the correct order. Points may need to be deselected first (see p 22).
Answers:

1 Angle ABC is 90°.

2 The angle is always 90°.

3 A perpendicular line through the centre of a circle and a chord bisects the chord.

SSM17: Investigating circles 2

Requirements: The Geometer's Sketchpad *File*: none
Aims:
■ use a dynamic geometry package to investigate properties of circles.
NC Reference: KS4 Ma3 2h(H), 2i(F)
Teacher preparation: make sure you know how to measure an angle in Geometer's Sketchpad (see page 22).
Assumed knowledge:
■ draw lines and measure angles in The Geometer's Sketchpad.
Support: SSM16
Starting the activity: explain the requirements of an investigation.
Hints: the labels on screen may be different to those on the activity sheet.
Answers:

1 Angle CAB is twice the size of angle CDB.

2 Angle EHF is always 90°.

3 The angles LNK and LMK are equal.

4 The sum of the opposite angles PSR and PQR is 180°.

5 The angles YVW and VXW are equal.

SSM18: Triangles, bisectors and circles

Requirements: The Geometer's Sketchpad *File*: none
Aims:
■ use a dynamic geometry package to look at the different circles obtained when the angles and sides of a triangle are bisected.
NC Reference: KS4 Ma3 4b(H), 4c(H), 4j(F)
Teacher preparation:
Assumed knowledge:
■ draw lines and add points, angle bisectors and perpendicular bisectors in The Geometer's Sketchpad.
Support: SSM17
Starting the activity: review the bisectors of angles and lines.
Hints: note that the centre point must be added after two bisectors have been drawn rather than all three.
Answers:

1 The circle just touches the sides of the triangle. The circle constructed is an incircle.

2 The circle passes exactly through the three vertices of the triangle. The circle constructed is a circumcircle.

3 The circle constructed is a circumcircle.

4 The circle constructed is a circumcircle.

SSM19: Investigating geometrical sequences

Requirements: calculator, Excel *File*: none
Aims:
■ use a spreadsheet to investigate the limit of a geometric sequence of circles and polygons
■ use trigonometry, angles in polygons in both degrees and radians and limits
■ create formulae.
NC Reference: KS4 Ma3 2f(H), 2g(H), 2h(H) (also Ma1)
Teacher preparation:
Assumed knowledge:
■ knowledge of trigonometry and geometry
■ enter trigonometric formulae in Excel
■ an understanding of radians
■ 3D trigonometry (for the extension).
Support: EX2
Starting the activity: introduce radians.
Hints: when you want to use a trigonometric ratio like cosine, the angle must be in radians and not degrees as Excel only works in radians.
π can be entered as pi().
Answers:

1 OD = 10 cos60 = 5 cm

2 **(a)** OE = 5 cm

   (b) Angle EOF = 45°

   (c) OF = 5 cos45 = 3.54 cm

Number of sides of the polygon	Angle x, in degrees	Angle x, in radians	Next radius	Starting radius	Ratio of starting radius to 'final' radius
3	60.0000	1.0472	5.0000	10	0.11505551
4	45.0000	0.7854	3.5355		
5	36.0000	0.6283	2.8603		
6	30.0000	0.5236	2.4771		
7	25.7143	0.4488	2.2318		
8	22.5000	0.3927	2.0619		
9	20.0000	0.3491	1.9376		
10	18.0000	0.3142	1.8427		
11	16.3636	0.2856	1.7681		
12	15.0000	0.2618	1.7078		
13	13.8462	0.2417	1.6582		
14	12.8571	0.2244	1.6166		
15	12.0000	0.2094	1.5813		
16	11.2500	0.1963	1.5509		
17	10.5882	0.1848	1.5245		
18	10.0000	0.1745	1.5014		
19	9.4737	0.1653	1.4809		
20	9.0000	0.1571	1.4626		

Changing the value of the starting radius has no effect on the ratio.

SSM20: Approximating a value for π

Requirements: Excel, scientific calculator
File: none
Aims:
■ using Pythagoras' theorem and area (sum of areas of trapeziums and triangle) to estimate a value for π using rough calculations and a spreadsheet.
NC Reference: KS4 Ma3 2f(H), 2h, 4d(H) (also Ma1)
Teacher preparation:
Assumed knowledge:
■ a good knowledge of a spreadsheet.
Support: EX2
Starting the activity: Q1–4 can be done in the classroom with the aid of a scientific calculator.
Hints: warn pupils not to drag down too many rows in one go. If printing, remind pupils to select an area of data to print, if it runs to more than one page.
Answers:

1 $y_1 = \sqrt{5^2 - 1^2} = 4.899$ to 3 d.p.

$y_2 = \sqrt{5^2 - 2^2} = 4.583$ to 3 d.p.

$y_3 = \sqrt{5^2 - 3^2} = 4$

$y_4 = \sqrt{5^2 - 4^2} = 3$

2 $T_1 \approx \frac{1}{2}(5 + 4.899) \times 1 \approx 4.590$

$T_2 \approx \frac{1}{2}(4.899 + 4.583) \times 1 \approx 4.741$

$T_3 \approx \frac{1}{2}(4.583 + 4) \times 1 \approx 4.292$

$T_4 \approx \frac{1}{2}(4 + 3) \times 1 \approx 3.5$

$T_5 \approx \frac{1}{2}(1 \times 3) \approx 1.5$

3 *Estimate area of Quadrant = 18.623*

4 *Estimate of $\pi = (18.623 \times 4) \div 5^2 = 2.980$*

5

x	y	Area	Radius	5
0	5	0.24999994	**Steps**	1000
0.005	4.9999975	0.24999962	**Step**	0.005
0.01	4.9999875	0.24999881		
0.015	4.999965	0.24999725		
0.02	4.999924999	0.24999469		
0.025	4.999862498	0.24998087		
0.03	4.999772495	0.24998556		
0.035	4.999649988	0.24998785		
0.04	4.999489974	0.24996944		
0.045	4.999287449	0.24995812		
0.05	4.999037407	0.24994431		
0.055	4.99873484	0.24992774		
0.06	4.998374736	0.24990616		
0.065	4.997952081	0.24988535		
0.07	4.997461856			
.	.	.		
.	.	.		
.	.	.		
4.99	0.316069613	0.001349051		
4.995	0.223550889	0.000558877		
	Total	19.63607072		
	Estimate	3.14177132		

6

x	y	Area	a	10
0	5.00000	0.04999999	**b**	5
0.01	5.00000	0.04999994		
0.02	4.99999	0.04999984	**Step**	100
0.03	4.99998	0.04999969	**Step**	0.01
0.04	4.99996	0.04999949		
0.05	4.99994	0.04999924		
0.06	4.99991	0.04999894		
0.07	4.99988	0.04999859		
0.08	4.99984	0.04999819		
0.09	4.99980	0.04999774		
0.1	4.99975	0.04999724		
0.11	4.99970	0.04999669		
0.12	4.99964	0.04999609		
0.13	4.99958			
.	.	.		
.	.	.		
.	.	.		
9.98	0.316069613	0.002698103		
9.99	0.223550889	0.001117759		
10	9.17215E-07	0.001117754		
	Total	39.2705611		
	Estimate	3.14164489		

7 The greater the number of trapeziums the more accurate the estimate of the value for π.

SSM21: Investigating trigonometric curves

Requirements: Omnigraph
File: none
Aims:
■ use a graphing package to investigate the connections between trigonometric curves, making use of transformations.
NC Reference: KS4 Ma3 3a, 3b (also Ma1 and Ma2 6g(H))
Teacher preparation:
Assumed knowledge:
■ enter trigonometric equations into Omnigraph.
Support: NA16
Starting the activity: review transformations and introduce any unknown terminology, e.g. maximum and minimum values.
Hints: remind pupils to use trigonometric scales and to rescale the axes as suggested.
Answers:

1 **(a)** They are 90° out of phase and $-1 \leq \sin x \leq 1$ and $-1 \leq \cos x \leq 1$.

(b) $\sin x$ crosses at 0°, ±180°, ±360°..... $\cos x$ crosses at ±90°, ±270°...

(c) $\sin x$ reaches its maximum at 90°±360° and minimum at 270° ±360°.... $\cos x$ reaches its maximum at 0°±360° and minimum at 180°±360°.

2 **(a)** $-\infty \leq \tan x \leq \infty$

(b) At 90°±180°

(c) At 0°±180°

3

Curve \ Angle	0°<x<90°	90°<x<180°	180°<x<270°	270°<x<360°
y=sin x	Positive	Positive	Negative	Negative
y=cos x	Positive	Negative	Negative	Positive
y=tan x	Positive	Negative	Positive	Negative

4 A reflection in y = 0

5 (a) $\cos(90°-x) = \sin x$

(b) $\sin(90°-x) = \cos x$

(c) $\sin(180°-x) = \sin x$

(d) $\cos(180°-x) = -\cos x$

(e) $\cos(360°+x) = \cos x$

(f) $\sin(360°+x) = \sin x$

(g) $\tan(180°-x) = -\tan x$

(h) $\tan(360°+x) = \tan x$

(i) $\sin(-x) = -\sin x x$

(j) $\cos(-x) = \cos x$

(k) $\tan(-x) = -\tan x$

6 (a) The transformation has the effect of squashing the curve horizontally. The value of *a* indicates the number of complete periods in the range 0° < x <360°.

(b) The transformation has the effect of stretching the curve horizontally. The value of *a* indicates what fraction of a complete period is in the range 0° < x <360°.

(c) The transformation has the effect of stretching the curve vertically. The value of *b* indicates the maximum and minimum values of the curve.

(d) The transformation has the effect of squashing the curve vertically. The value of *b* indicates the maximum and minimum values of the curve.

(e) The transformation has the effect of squashing the curve vertically by a factor of 2 and horizontally by a factor of 3.

(f) The transformation has the effect of stretching the curve vertically by a factor of 2 and horizontally by a factor of 3.

HD1: Mathematics through the ages

Requirements: internet
File: none
Aims:
- use a given website to find information about famous mathematicians to help them solve a crossword puzzle.

NC Reference: KS3 Ma4 1a, 1b, 1c, 1d, 1g, 4b

Teacher preparation: check that there is access to the internet; website addresses.
Assumed knowledge:
- none

Support:
Starting the activity: ask pupils to name any famous mathematicians.
Hints: remind pupils to stay on the given websites.
Answers:

```
P A S C A L . . . . . . C .
Y . A . . M . . . E . . A .
T . M . . A . . . S . . R .
H . B . . C . . . C . . R .
A . R U D O L F F . H . O .
G . I . . A . I . E . . L .
O . D . . U . B E R N O U L L I
R . G . . R . O . . . . . . .
A H E R O . I . N . D . . . .
S . A . . . N . A . O H M . F
. . N . . . . . C . D . . R .
. . N . . . . . C . G . . E .
I S A A C . . . E I N S T E I N
. H . . . . . . . . O . . C .
. . . . . . . . . . N . . H .
```

HD2: A day out in London

Requirements: internet, Excel
File: none
Aims:
- use the internet to find out information for a day out and makes use of the 24-hour clock and the calendar.
- extension: capture images and produce a slide show of their day.
- use a spreadsheet to record their expenditure.

NC Reference: KS3 Ma4 1a, 1b, 1c, 1d, 3b, 5a, 5e
Teacher preparation: check that there is access to the internet; website addresses.
Assumed knowledge:
- enter data and formulae in Excel.

Support: EX2
Starting the activity: discuss what is required for a family day out.
Hints: remind pupils to stay on the given websites. Remind pupils to keep a record of their expenses and activities.
Answers: none

HD3: Flights

Requirements: internet   *File*: none
Aims:
- use information from a given website to find out flight times.
- use the 12- and 24-hour clock and the calendar.

NC Reference: KS3 Ma4 1a, 1d, 3b, 5a, 5e
Teacher preparation: check there there is access to the internet; website addresses.

Assumed knowledge:
- 24-hour clock.

Support:
Starting the activity: ask pupils to name some cities around the world.
Hints: remind pupils to stay on the given websites. As an extension pupils could use a spreadsheet to create a bar chart to display the duration of the journeys.
Answers: none

HD4: Looking at data 1

Requirements: Excel
File: Soccer League Tables.xls
Aims:
- use a spreadsheet to look at formulae, mean averages, sorting, analysing data, scatter graphs and correlation.
- modelling changes to league positions by changing the number of points for a win.

NC Reference: KS3 Ma4 1f, 3b, 4a, 4e, 4j, 5f
Teacher preparation: check that pupils can access the file.
Assumed knowledge:
- enter formula into Excel
- sort data, produce scatter graphs and look for correlation.

Support: EX2, EX3
Starting the activity: take the pupils through Q1.
Hints: remind pupils to keep a record of their findings.

Answers:

1 (a)–(c)

Club	Home Wins	...	Away Goals conceded	Total Points
Nottingham F.	18	...	22	94
Middlesborough	17	...	29	91
...	...	...	...	...
Manchester City	6	...	31	48
Stoke City	8	...	34	46
Reading	8	...	47	42

Club	Home Wins	...	Away Goals conceded	Total Points	Goal Difference
Nottingham F.	18	...	22	94	40
Middlesborough	17	...	29	91	36
...	...	...	...	...	...
Manchester City	6	...	31	48	−1
Stoke City	8	...	34	46	−30
Reading	8	...	47	42	−39

Club	Home Wins	...	Away Goals conceded	Total Points	Goal Difference
Nottingham F.	18	...	22	94	40
Middlesborough	17	...	29	91	36
...	...	...	...	...	...
Manchester City	6	...	31	48	−1
Stoke City	8	...	34	46	−30
Reading	8	...	47	42	−39

1 (d) The three teams that were relegated were Manchester City, Stoke City and Reading.

3 The total for the goal difference should be zero.

4

Club	Total Points	Goal Difference	Total Goals scored
Nottingham F.	94	40	82
Middlesborough	91	36	77
Sunderland	90	36	86
Charlton	88	31	80
Ipswich	83	34	77

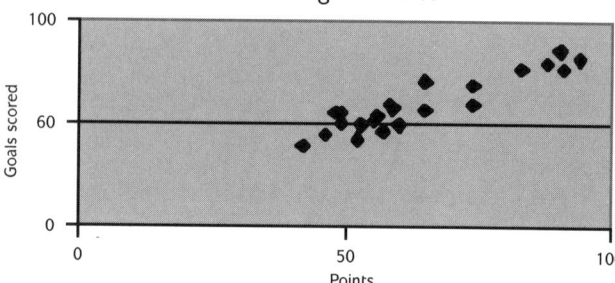
Points versus goals scored

There is a positive correlation between the number of points gained and the number of goals scored.

2 (a)

Club	Home Wins	Home Draws	Home Defeats	Home Goals scored	Home Goals conceded	Away Wins	Away Draws	Away Defeats	Away Goals scored	Away Goals conceded
Nottingham F.	18	2	3	52	20	10	8	5	30	22
Middlesborough	17	4	2	51	12	10	6	7	26	29
...	...	...	...	...	...	...	...	...	...	...
Manchester City	6	6	11	28	26	6	6	11	28	31
Stoke City	8	5	10	30	40	3	8	12	14	34
Reading	8	4	11	27	31	3	5	15	12	47
Mean	10.9	6	6	34.8	24	6	6.1	10.9	24	34.8

(b) Some of these results are the same because a home win for one team is an away defeat for another.

5

Club	Total Points	Goal Difference	Total Goals scored	Total Goals conceded
Nottingham F.	94	40	82	42
Middlesborough	91	36	77	41
Sunderland	90	36	86	50
Charlton	88	31	80	49
Ipswich	83	34	77	43

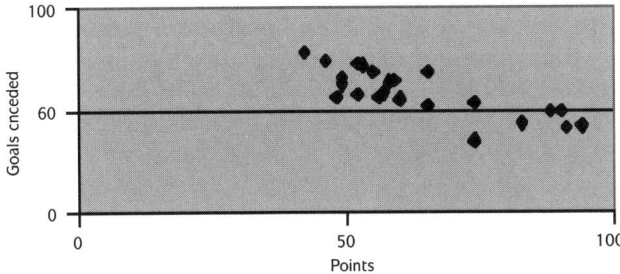

Points versus goals conceded

There is a negative correlation between the number of points gained and the number of goals scored.

6

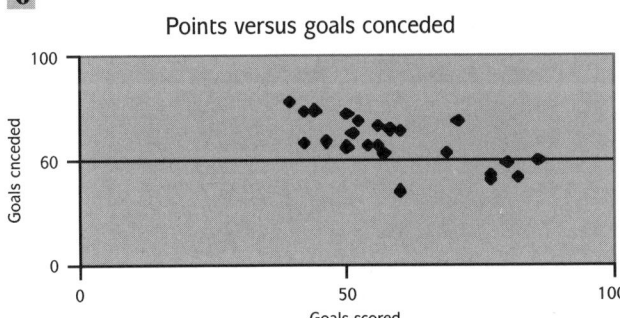

Points versus goals conceded

There is a low negative correlation between goals scored and goals conceded.

7 (b)

Club	Home Wins	Home Draws	...	Total Points
Nottingham F.	18	2	...	66
Middlesborough	17	4	...	64
...	...	...	...	...
Manchester City	6	6	...	36
Port Vale	7	6	...	36
Portsmouth	8	6	...	36
Stoke City	8	4	...	35
Reading	8	4	...	31

7 (c)

Club	Home Wins	Home Draws	...	Total Points	Goal Difference
Nottingham F.	18	2	...	66	40
Middlesborough	17	4	...	64	36
...	...	...	...	...	...
Manchester City	6	6	...	36	−1
Port Vale	7	6	...	36	−10
Portsmouth	8	6	...	36	−12
Stoke City	8	4	...	35	−30
Reading	8	4	...	31	−39

7 **(d)** Portsmouth, Stoke City and Reading would have been relegated. Manchester City and Portsmouth were most affected since they have swapped relegation positions.

HD5: Looking at data 2

Requirements: Excel, access to a printer
File: olympics.xls
Aims:
■ use a spreadsheet to describe social trends, interpret data, work out percentages and make predictions.
NC Reference: KS3 Ma4 1f, 5b, 5c, 5d, 5f
Teacher preparation: check that pupils can access the file.
Assumed knowledge:
■ produce scatter graphs and drag data in Excel
■ enter a formula to calculate percentages.
Support: EX2, EX3
Starting the activity: do Q1 orally.
Hints: remind pupils to place charts as new sheets.
Answers:

1 **(a)** Due to the games being held in the USA and travelling from Europe was difficult.
(b) 1916, 1940 and 1944.
(c) Due to the two world wars.
(d) Large countries, much larger talent base.
(e) Did not really take part, cold war introduced a great deal of rivalry.
(f) The USA boycotted the Moscow Olympic Games and The USSR boycotted the Los Angeles Olympic Games.

2 **(a)**

Year	City hosting the Games	Number of male competitors	Number of female competitors	Mens 400m winning time in seconds
1896	Athens	200	0	54.2
1900	Paris	1225	19	49.4
1904	St.Louis	687	6	49.2
.	.	.	.	.
.	.	.	.	.
.	.	.	.	.
2024				40.0352964
2028				39.6750593

Men's 400m drops below 40s by 2028.

(b)

Year	City hosting the Games	Number of male competitors	Number of female competitors	Mens 400m winning time in seconds	Womens 400m winning time in seconds
1896	Athens	200	0	54.2	
1900	Paris	1225	19	49.4	
⋮	⋮	⋮	⋮	⋮	⋮
2056					40.3
2060					39.8

Womens's 400m drops below 40s by 2060.

3 **(a)** The percentage of women competitors fell in 1904, 1932, 1936 and 1940.

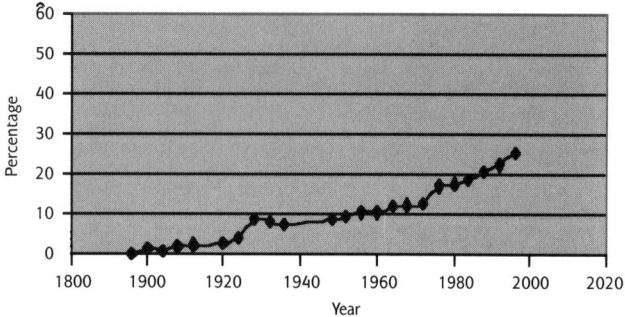

Percentage of Women Competitors

(c) Percentage of female competitors will reach 50% by approximately 2030.

4

Year	City hosting the games	Number of male competitors	Number of female competitors	Mens 400m winning time in seconds	Womens 400m winning time in seconds	Mens High Jump winning height in inches
1896	Athens	200	0	54.2		71.25
1900	Paris	1225	19	49.4		74.74
⋮	⋮	⋮	⋮	⋮	⋮	⋮
2056				41.80		107.49
2060				41.30		108.57

Men's high jump will reach 9ft by 2048.

HD6: Business trip

Requirements: internet
File: none
Aims:
■ use the internet to plan a business trip
■ use the 24-hour clock.
■ use given websites to do currency conversions and calculate distances travelled.
NC Reference: KS3 Ma4 1a, 1d, 3b, 5a, 5e (also Ma2 3o, 3q, 4a)
Teacher preparation: check that there is access to the internet; website addresses.

Assumed knowledge:
■ none
Support:
Starting the activity: tell pupils to enter the airports as they are in the table on the activity sheet.
Hints: remind pupils to stay on the given websites.
Answers: none

HD7: Cumulative frequency

Requirements: Excel, access to a printer
File: none
Aims:
■ use a spreadsheet to estimate medians, upper and lower quartiles and interquartile ranges from frequency curves.
NC Reference: KS3 Ma4 4a, 4b, 5b
Teacher preparation:
Assumed knowledge:
■ enter formulae into Excel
■ produce a cumulative frequency curve with the aid of a scatter graph in Excel.
Support: EX2, EX3
Starting the activity: review the terminology used in the activity.
Hints: remember to use the points at the upper end of the range when plotting points from a grouped frequency table. Remind pupils to add labels to the axes and a title to the curve. Place the chart as a new sheet.
Answers:

1 **(a)**

Speed	Frequency	Cumulative Frequency
40	2	2
50	3	5
60	14	19
70	16	35
80	33	68
90	47	115
100	52	167
110	22	189
120	11	200

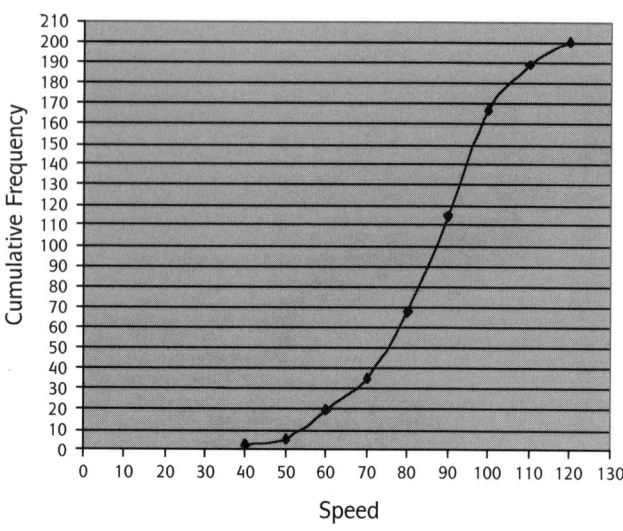

(c) (i) Estimate of Speed ≈ 88km⁻¹

   (ii) Estimate of Interquartile Range ≈ 21
(Upper Quartile ≈ 96, Lower Quartile ≈ 75)

   (iii) Estimate of Number of Vehicles travelling more than 85 km⁻¹ ≈ 110

   (iv) Estimate of Number of Vehicles travelling less than 65 km⁻¹ ≈ 26

2 **(b)** Time interval 46-60 has the modal number of goals.

(c)

Time, in minutes, from	Time, in minutes, to	Frequency	Cumulative frequency
1	15	8	8
16	30	9	17
31	45	17	34
46	60	23	57
61	75	13	70
76	90	12	82

(e) The total number of goals is 82

(f)

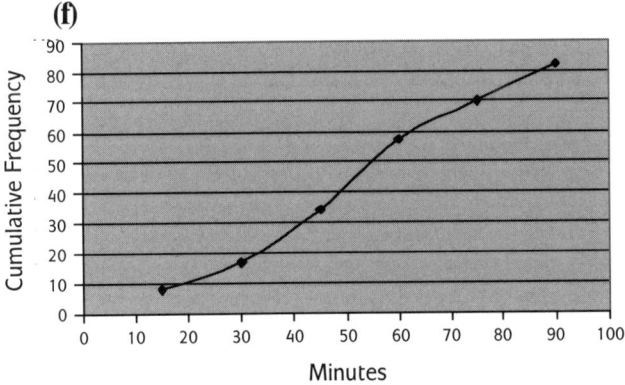

(g) (i) Estimate of Number of Goals scored after 55 minutes ≈ 32

   (ii) Estimate of Number of Goals scored before 75th minute ≈ 70

HD8: Quartiles, averages and ranges

Requirements: Excel
File: houseprice.xls
Aims:
- use a spreadsheet containing house price data to calculate lower and upper quartiles, interquartile ranges, means and medians.

NC Reference: KS3 Ma4 1a (iii), 4a, 4b, 5d; KS4 Ma4 4a, 4b(F), 4e(H), 4j(H), 5d
Teacher preparation: check that pupils can access the file. If available, obtain a set of graphical calculators to assist with Q1(j). SPSS or Autograph will also display box and whisker plots.
Assumed knowledge:
- enter formulae into Excel.
Support: EX2

Starting the activity: discuss the reasons behind the range of given data.
Hints: =median(B2:B21) will also calculate the median.
Answers:

1

Region	Price in pounds
South Humberside	£ 44,800.00
Dyfed	£ 48,400.00
South Yorkshire	£ 50,300.00
Clwyd	£ 51,700.00
Cleveland	£ 52,300.00
County Armagh	£ 52,950.00
North Humberside	£ 53,700.00
Lincolnshire	£ 53,900.00
Mid Glamorgan	£ 54,600.00
County Durham	£ 55,750.00
Hampshire	£ 116,250.00
Kent	£ 119,700.00
Buckinghamshire	£ 120,050.00
East Sussex	£ 132,700.00
West Sussex	£ 137,250.00
Oxfordshire	£ 137,600.00
Hertfordshire	£ 158,400.00
Berkshire	£ 164,150.00
Surrey	£ 179,250.00
Greater London	£ 210,100.00
Range	**£ 165,300.00**
Mean	**£ 99,692.50**
Lower Quartile	**£ 52,787.50**
Median	**£ 86,000.00**
Upper Quartile	**£ 137,337.50**
Interquartile Range	**£ 84,550.00**

(h) There is such a large difference because the data is poorly spread. There are no values between £55,570 and £116,250.

(i) They would generally choose the value which looks the most favourable. In this case they may choose the mean as it is the lower value and would indicate less of a divide.

HD9: Looking at data 3

Requirements: Excel, access to a printer
File: transplants.xls
Aims:
- use a spreadsheet to analyse data on transplant procedures and teenage pregnancies, produce scatter graphs, look at trends and lines of best fit.

NC Reference: KS3 Ma4 4a, 5b, 5c, 5g
Teacher preparation: check that pupils can access the file; pupils can access a printer.
Assumed knowledge:
- produce scatter graphs in Excel
- dragging down data in Excel to produce trends.

Starting the activity:
Hints: remind pupils to place their charts as a new sheet.
Answers:

1 (b)

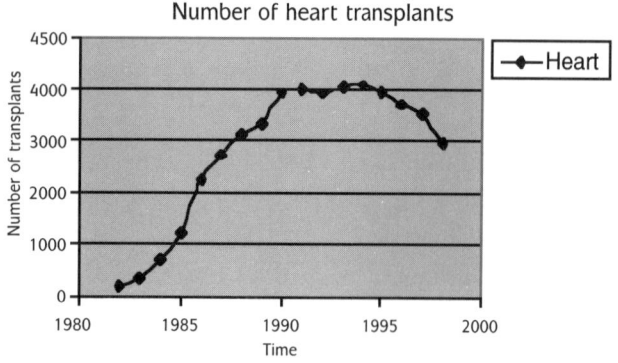

Number of heart transplants

(b)

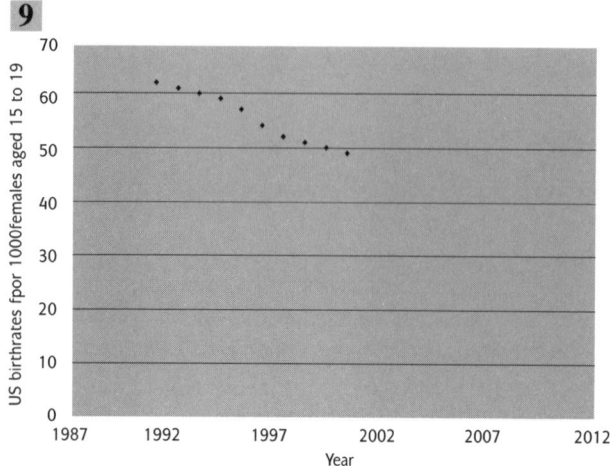

Number of transplants

2 First decrease in 1992.

3 The trend was then to increase slightly through 1993-1994 and then to steadily decrease.

4 Single lung transplants overtook heart-lung transplants in 1990

5 The overall trend of all the different types of transplant is a gradual increase until 1994 after which they level out of decline.

6 Shows a decreasing trend.

9

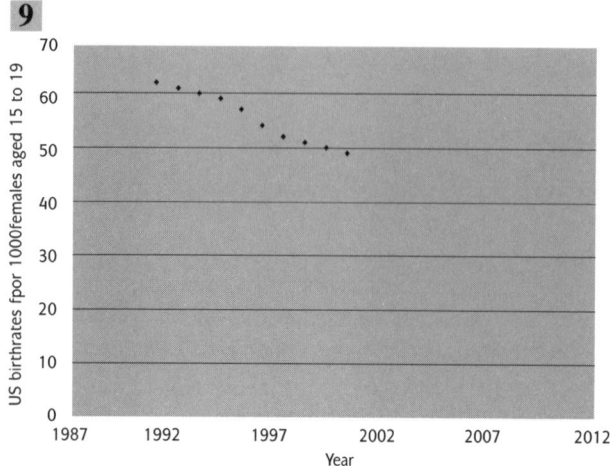

10 2012.

12 Produces same date, 2012.

HD10: Probability models

Requirements: Excel
File: onedice.xls, twodice.xls, lottery.xls
Aims:

■ use a given spreadsheet to compare experimental probabilities with theoretical probabilities.
NC Reference: KS3 Ma4 5a, 5c, 5h, 5j; KS4 Ma4 5h
Teacher preparation: Before using these templates it is important to make sure your version of Excel is able to run the probability models. To do this:

   ■ launch Excel
   ■ click Tools then Add-Ins...
   ■ make sure the following two are ticked

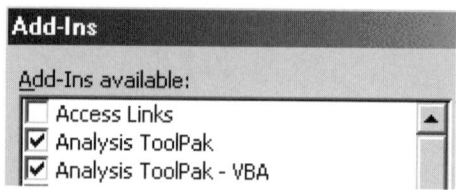

   ■ and click the OK button.
Check that pupils can access the files.
Assumed knowledge:

■ none.
Support:
Starting the activity: discuss the terms theoretical and experimental probability and why they may differ.
Hints:
Answers:

1 (a) Theoretical probability = 1/6.
   (b) $1/6 = 0.1667$ (4dp).
   (c) Expected 20 2s.
   (d) Expected 120 2s.

2 The values will be similar but unlikely to be exactly the same.

3 The values will become increasingly more accurate.

4 Yes, it should be true that the more rolls the more accurate the results.

5

Totals for two dice	2	3	4	5	6	7	8	9	10	11	12
Probabilities	$\frac{1}{36}$	$\frac{2}{36}$	$\frac{3}{36}$	$\frac{4}{36}$	$\frac{5}{36}$	$\frac{6}{36}$	$\frac{5}{36}$	$\frac{4}{36}$	$\frac{3}{36}$	$\frac{2}{36}$	$\frac{1}{36}$

6

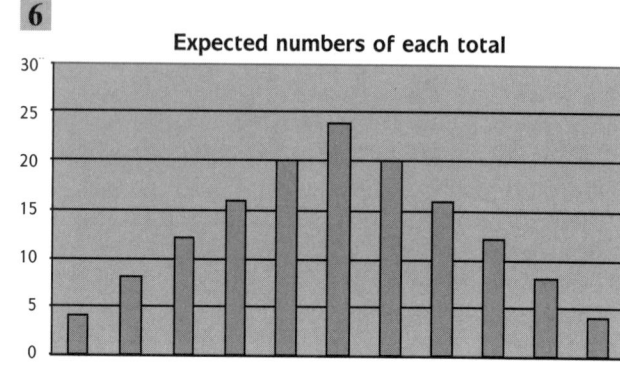

Expected numbers of each total

8 From the model around 45% of draws have consecutive numbers. You would obtain consecutive numbers almost every other draw.

9 It may well be a good idea to choose consecutive numbers but choosing which ones still causes a problem. The probability of winning is still very small. If there were no consecutive numbers the previous week why not choose them next time!

HD11: Looking at data 4

Requirements: Excel, internet
File: euro.xls
Aims:
■ use a spreadsheet to look at totals, mean, median and mode, scatter graphs, correlation
■ extension: entering and interpreting data acquired from given websites.
NC Reference: KS4 Ma4 3a, 3b(H), 3c(F), 4a, 4e(H), 4g(F), 4j(H), 5f
Teacher preparation: check that pupils can access the file; that there is access to the internet; website addresses.
Assumed knowledge:
■ enter formulae into a given Excel file
■ produce scatter graphs in Excel.
Support: EX2, EX3
Starting the activity:
Hints: remind pupils to place charts as new sheets; to use the Ctrl key to highlight non-adjacent columns.
Answers:

1

Team	Goals scored	Goals conceded	Points
Belgium	3	5	3
Czech republic	2	3	3
Denmark	0	8	0
England	5	6	3
France	7	4	6
Germany	1	5	1
Holland	7	2	9
Italy	6	2	9
Norway	1	1	4
Portugal	7	2	9
Romania	4	4	4
Slovenia	4	5	2
Spain	6	5	6
Sweden	2	4	1
Turkey	3	2	4
Yugoslavia	7	7	4
	65	**65**	
Mean	4.0625		
Median	4		
Mode	7		

2 (a) The modal number of points is 4, gained by Norway, Romania, Turkey and Yugolslavia.

(b) The median number of points is 4.

3 (a) The mean average number of goals scored is 4.0625

(b) The median average number of goals scored is 4.

(c) The modal number of goals scored is 7.

(d) The organisers of the Euro 2000 would use the mode to suggest a lot of attacking football. Goalkeepers would prefer the other two averages.

4 (b) Although fairly scattered there does tend to be a slight positive correlation between goals scored and points, also a slight negative correlation between goals conceded and points.

5

A scatter graph displaying the number of goals scored plotted against the number of points gained

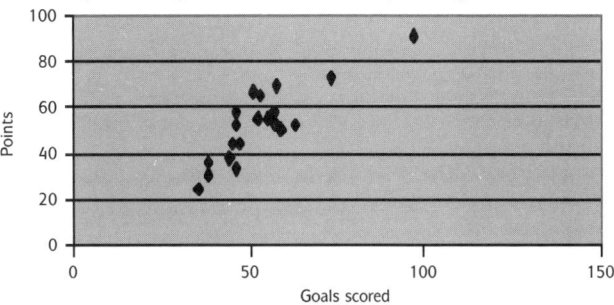

A scatter graph displaying the number of goals conceded plotted against the number of points gained

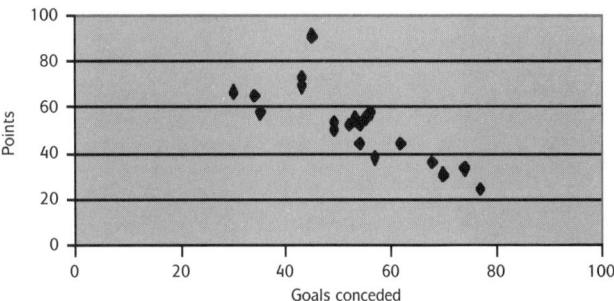

(b) We can clearly see here a positive correlation between the goals scored and the points gained and a negative correlation between the goals conceded and the points gained.

HD12: Estimating the mean

Requirements: Excel *File*: none
Aims:
■ use a spreadsheet to estimate the mean from a grouped frequency table.
NC Reference: KS4 Ma4 4e(H), 4g(F), 4j(H)
Teacher preparation:
Assumed knowledge:
■ enter formulae into Excel.
Support: EX2
Starting the activity: discuss the difference between grouped and ungrouped data and why the mean has to be estimated.
Hints: remind pupils to take care when using brackets in formulae.

Answers:

1

Marbles from	Marbles to	Frequency of guess	Midpoint of Range	Frequency x midpoint
251	300	12	275.5	3306
301	350	24	325.5	7812
351	400	88	375.5	33044
401	450	165	425.5	70207.5
451	500	97	475.5	46123.5
501	550	74	525.5	38887
551	600	33	575.5	18991.5
601	650	7	625.5	4378.5
		500		**222750**
				445.5

(**f**) (i) 401-450 contains the median number of marbles.

(ii) 401-450 is the modal class.

(**g**) Discussion

2 (a)

Pulse Rate from	Pulse Rate to	Frequency of guess	Midpoint of Range	Frequency x midpoint
30	34	1	32	32
35	39	3	37	111
40	44	7	42	294
45	49	24	47	1128
50	54	45	52	2340
55	59	62	57	3534
60	64	79	62	4898
65	69	83	67	5561
70	74	92	72	6624
75	79	70	77	5390
80	84	42	82	3444
85	89	31	87	2697
		539		**36053**
				66.8868

(**b**) (i) The modal pulse rate is in the group 70–74.

(ii) The median pulse rate is in the group 65–69

(**c**) Discussion

HD13: Investigating standard deviation

Requirements: Excel

File: none

Aims:

■ use a spreadsheet to calculate the standard deviation of sets of data.

NC Reference: Not required for GCSE Maths starting in Sep 2001. However, it may be useful in GCSE Statistics coursework from 2003.

Teacher preparation:

Assumed knowledge:

■ enter formulae into Excel.

Support: EX2

Starting the activity: discuss methods of interrogating data from a coordinate perspective.

Hints:

Answers:

1

Height in cm	Height in mm	Increased Height
58	580	595
53	530	545
52	520	535
47	470	485
61	610	625
47	470	485
64	640	655
64	640	655
58	580	595
44	440	455
62	620	635
52	520	535
6.9783214	69.783214	69.783214

(**c**) Multiplying the data by 10 changes the standard deviation by multiplying it by the same amount.

(**e**) Adding 15mm to the data does not change the standard deviation.

2

2a	2c	2e	2g
9	18	12	28
14	28	17	43
8	16	11	25
13	26	16	40
12	24	15	37
15	30	18	46
7	14	10	22
11	22	14	34
13	26	16	40
19	38	22	58
16	32	19	49
15	30	18	46
10	20	13	31
11	22	14	34
6	12	9	19
14	28	17	43
16	32	19	49
12	24	15	37
13	26	16	40
15	30	18	46
3.300319	6.6006379	3.300319	9.9009569

(**b**) Multiplying the data by 2 changes the standard deviation by multiplying it by the same amount.

(**d**) Adding 3 to the data does not change the standard deviation.

(**f**) Multiplying the data by 3 and adding 1 only changes the standard deviation by multiplying it by 3.

4 Activity sheets

Activity sheet EX1

Getting started in Excel 1

This activity sheet shows you how to format and sort numbers and text in Excel. Open the spreadsheet student1.xls. Part of it is shown below:

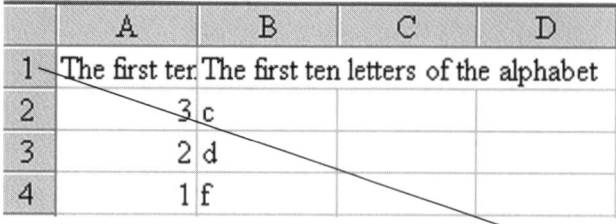

1 To improve the layout of the headings:
- Click on the row number 1 to highlight the entire row:

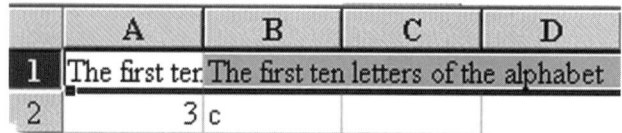

- On the main toolbar click on Format, Cells...then the Alignment tab..

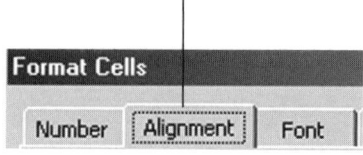

...and click here to add a tick...

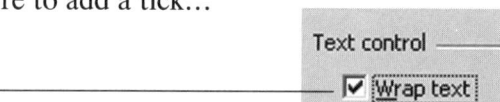

...and then click the OK button to produce the following column headings:

	A	B
1	The first ten whole numbers	The first ten letters of the alphabet
2	3	c

(continued)

2 In Excel the numbers in the cells are *right justified* [at the right of the cell] and the letters are *left justified*. To improve the layout it is a good idea to centre all of the data. To do this for the two columns A and B:

- Click on the column letter A, hold down the Ctrl key on the keyboard and click on the column heading B to highlight both columns
- Click the *Center* [US spelling!] button

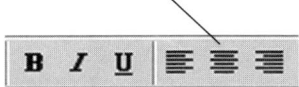

on the Format toolbar to produce:

	A	B
1	The first ten whole numbers	The first ten letters of the alphabet
2	3	c
3	2	d
4	1	f

Cell A2

3 To sort the numbers into *ascending* order:

- Click on Cell A2 and drag the mouse to highlight the rest of the cells down to cell A11
- On the main toolbar click the *Sort Ascending* button

4 To sort the letters into descending order:

- Click on Cell B2 and drag the mouse to highlight the rest of the cells down to cell B11
- On the main toolbar click the *Sort Descending* button

(a) Sort column C in ascending order. What appears in cell C1?
(b) Sort column D in descending order. What appears in cell D10?

5 You can further change the way your spreadsheet looks by... highlighting all or parts of your data and using Format, Cells... and the *Font*, *Border* and *Patterns* tabs to add colour and borders. Try changing the appearance of your worksheet.

Activity sheet EX2

Getting started in Excel 2

This activity sheet shows you how to enter and use formulae to generate sequences in Excel. Open the spreadsheet student2.xls. Part of it is shown below:

	A	B	C
1	The first ten whole numbers	The first ten odd numbers	The first ten even numbers
2	1	1	2
3			

1 Click on cell A3 and enter the formula =A2+1 ————————— This formula adds 1 to the number in cell A2

2 Click on cell A3 and drag the square at the bottom right of the cell down to cell A11 to produce the first 10 whole numbers.

	A
1	The first ten whole numbers
2	1
3	2

———————— Drag this square down to cell A11

3 Click on cell A4 and note the formula in the formula bar.

A4 ▼ = =A3+1

4 What formula do you think is contained in cell A11? Click on the cell to check your answer in the formula bar.

5 Each odd number is 2 more than the previous one so enter the formula =B2+2 in cell B3 and drag its black square down to cell B11 to produce the first ten odd numbers.

6 What formula do you think is contained in cell B8? Click on the cell to check your answer.

7 Use a formula in cell C3 to generate the first 10 even numbers.

8 Use the formula =D2+5 in cell D3 to generate the first 10 multiples of 5 in column D.

(continued)

EX2 – (continued)

9 Use a formula in cell E3 to generate the first 10 multiples of 10 in column E.

10 The Fibonacci sequence begins:

1, 1, 2, 3, 5, 8, ...

where each term after the second is the sum of the previous two.

Write down the next 4 Fibonacci numbers:

Use the formula =F2+F3 in cell F4 and drag it down to cell F10 to check your answers.

What will the formula for cell F9 be?

Click on cell F9 to check your answer.

11 Click on the 'Multiplication table' tab at the bottom of the worksheet.

◄◄ ► ►◄ \ Sequences \ **Multiplication table** /

Enter the formula =A2*B1 in cell B2 and drag it down to cell B13.

Click on cell B3 and then B4 and note in the formula bar how the dollar signs in the formula have stopped B1 becoming B2, then B3 etc. after you dragged down.

Enter the formula =A2*C1 in cell C2 and drag it down to cell C13.

Use a formula in cells D2 to M2 to complete the multiplication table.

Print out a copy of your multiplication table.

> The multiplication sign in Excel is: *
>
> The division sign in Excel is: /

> Hints:
> In column A each term is twice the previous one.
> In column B each term is one less than the previous one.
> In column C each term is one tenth of the previous one.
> In column D each term is three times the previous one.
> In column E each term is five less than the previous one.

12 Click on the 'More sequences' tab and use row 2 to generate the first 15 terms of these sequences:

	A	B	C	D	E
1	2	20	100000	1	5
2	4	19	10000	3	0
3	8	18	1000	9	-5
4	16	17	100	27	-10
5	32	16	10	81	-15
6	64	15	1	243	-20

13 Extension: Use columns F and G to generate the first 15 terms of the following two sequences:

2, 4, 10, 28,….. and 3, 11, 27, 59,…

Activity sheet EX3

Getting started in Excel 3

This activity sheet shows you how to *display data as bar, pie and line graphs* in Excel.

The data below was obtained from the website **http://weather.yahoo.com/forecast/Folkestone_UK_f.html** on Sunday the 9th September 2001:

5 Day Forecast

Today	Tomorrow	Tue	Wed	Thu
Mostly Cloudy	Partly Cloudy	Mostly Cloudy	Mostly Cloudy	Showers
High: **57**	High: **61**	High: **59**	High: **63**	High: **68**
Low: **46**	Low: **46**	Low: **48**	Low: **52**	Low: **55**

1 Open the spreadsheet student3.xls containing these High and Low temperatures. Part of it is shown below:

	A	B	C
	Day of the week	Highest temperature	Lowest temperature
1			
2	Sunday	57	46
3	Monday	61	46
4	Tuesday	59	48
5	Wednesday	63	52
6	Thursday	68	55

2 Click on cell A1 and drag the mouse to highlight all the data in columns A, B and C from row 1 down to row 6.

3 Click the chart wizard button on the main toolbar...

...and select at stage one of the wizard:

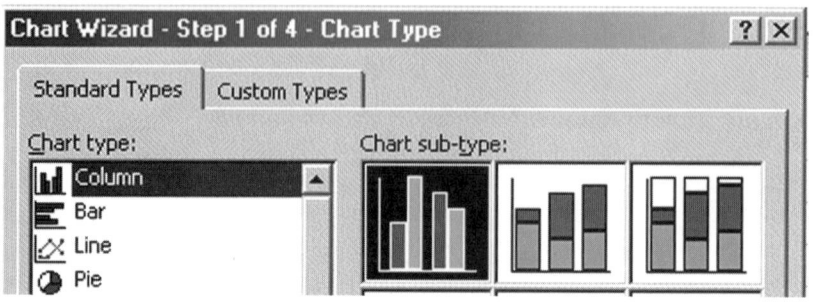

Note: In Excel what we call a bar chart is called a column chart!

It is a good idea to press and hold this button to check the chart:

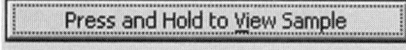

(continued)

4 Click the 'Next >' button to see your chart at step 2 of the wizard and if it looks correct [it should!], click the 'Next >' button again to reveal step 3.

5 Enter:

■ A bar chart comparing the highest and lowest temperatures in a 5 day period

in the 'Chart title box';

■ Day of the week

in the 'Category (X) axis' box;

■ Temperature in degrees Celsius

in the Value (Y) axis box and click the 'Next >' button again to reveal step 4.

6 Finally choose to Place your chart as a new sheet:

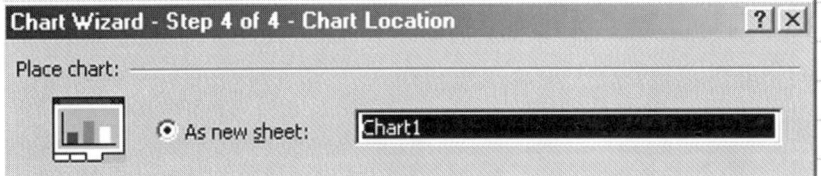

and click the 'Finish' button to produce your bar chart:

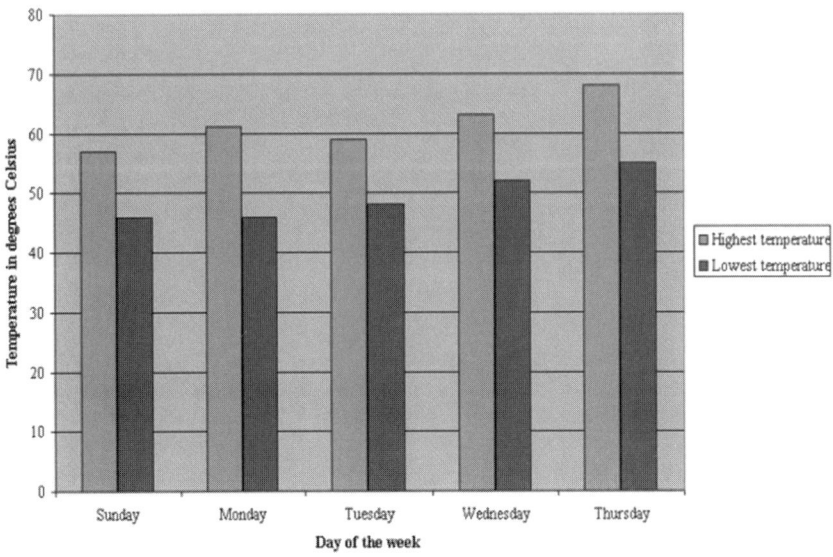

Print out a copy of your bar chart.

7 Click the 'Sales figures' tab at the bottom of your spreadsheet.

The figures show the quarterly sales for the '1729 Mathematics Software' company.

(continued)

Highlight all the data, including the headings, and produce a line graph of the sales.

Remember to
- add suitable labels to your graphs
- place the chart as a new sheet
- print out a copy of your graph

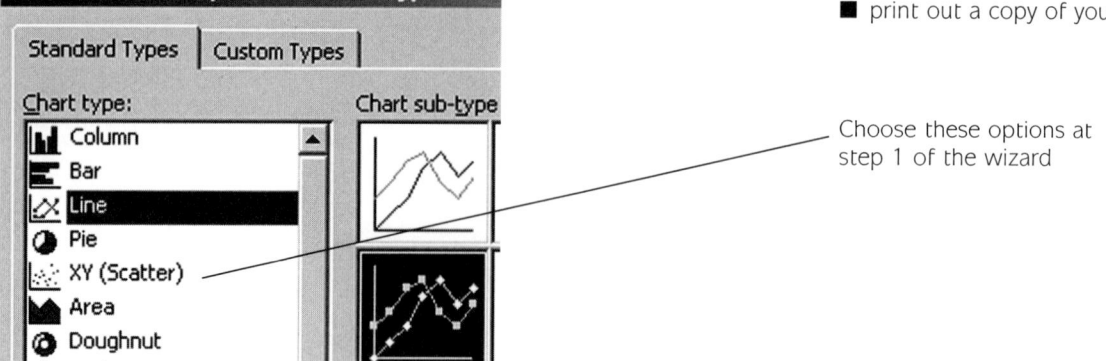

Choose these options at step 1 of the wizard

8 Click the 'Pulse rates' tab at the bottom of your spreadsheet. The data shows the resting pulse rate of 5 students.

Create a pie chart to display the data.

9 Extension: Collect some data from the internet, newspapers or questionnaires and create some suitable charts to display your data.
Some suggested websites:

http://www.teamtalk.com/ ———————————— Choose a soccer or rugby club and find the *players* link to gain some data on age, height and weight.

http://weather.yahoo.com/

http://www.theaa.com/ ———————————— Use the *running costs* to gain all sorts of data on cars.

Activity sheet NA1

My new room

This activity sheet shows you how you can use a spreadsheet to model a situation. Before starting the activity, launch Excel and open the spreadsheet mynewroom.xls.

	A	B	C	D	E	F	G	H
1	My New Room							
2	Wallpaper	Cost Per Roll	Quantity	Total	Paint	Cost	Quantity	Total
3	Beige Leaf	£ 5.99	2	£ 11.98	Brilliant White Gloss	£ 6.99	1	£ 6.99
4	Sea Blue Swirl	£ 7.99	0	£ -	Blue Gloss	£ 7.99	0	£ -
5	Natural Wash	£ 9.99	3	£ 29.97	Dreamy Cool Green	£ 10.99	0	£ -
6	Morning Yellow Stripe	£ 9.99	0	£ -	Warm Orange	£ 10.99	0	£ -
7	Green Grass Stripe	£ 14.99	0	£ -	Germa Purple	£ 10.99	0	£ -
8			5	£ 41.95			1	£ 6.99
9								
10	Beds	Cost	Quantity	Total	Lamps	Cost	Quantity	Total
11	Bunk with Desk	£ 229.00	1	£ 229.00	Simple Desk	£ 9.99	1	£ 9.99
12	Convoy Single	£ 219.00	0	£ -	Adjustable Desk	£ 24.99	0	£ -
13	Metal Frame Single	£ 239.00	0	£ -	Hanging	£ 34.99	1	£ 34.99
14	Hideaway Guest Bed	£ 279.00	0	£ -	Candle Stick	£ 39.99	0	£ -
15	Single Divan	£ 249.00	0	£ -	Aluminium Pendant	£ 24.99	0	£ -
16			1	£ 229.00			2	£ 44.98
17								
18	Computer	Cost	Quantity	Total	Desk	Cost	Quantity	Total
19	Dell Internet Ready 500Mhz	£ 599.00	1	£ 599.00	Metal Frame	£ 39.99	1	£ 39.99
20	Tiny Internet Ready 550Mhz	£ 659.00	0	£ -	Corner Desk	£ 159.00	0	£ -
21	PC Word Internet Ready 650Mhz	£ 699.00	0	£ -	Dark Brown Desk	£ 259.00	0	£ -
22	Gateway Laptop	£ 1,099.00	0	£ -	PC Station	£ 59.99	0	£ -
23	Eton Athalon 850Mhz	£ 999.00	0	£ -	Home Office	£ 179.00	0	£ -
24			1	£ 599.00			1	£ 39.99
25								
26						Total:	£	961.91
27						Budget:	£	1,500.00
28						**Balance:**		**£538.09**

You have been given a budget of £1500 to redecorate your new room.

You need to buy:

1 bed, 5 rolls of wallpaper, 2 lamps, 1 can of paint, 1 computer and 1 desk.

2	Wallpaper	Cost Per Roll	Quantity	T
3	Beige Leaf	£ 5.99	2	£
4	Sea Blue Swirl	£ 7.99	0	£
5	Natural Wash	£ 9.99	3	£

1 Amend the entries and select the **most** expensive items and write down how much over budget you are:

To select an item, simply change the quantity.

£ _____

2 Select the **least** expensive items and write down how much under budget you are:

£ _____

3 By altering the quantities of the different items, see how close you can get to the £1500 budget.
Enter your final total:

£ _____

4 At the January sales all items were reduced by 15%. Amend columns B and F on your worksheet to show the new prices and make changes to see if you can now purchase more expensive items.

To reduce a price by 15% multiply it by 0.85

Activity sheet NA2

Unit conversions

This activity sheet shows you how to use a spreadsheet to convert metric and imperial units, temperature and currency exchange rates. Before starting the activities, launch Excel.

1 To convert the metric measure of weight, kilograms, kg, into its imperial equivalent, pounds, lb, you can multiply by 2.2.

In column A generate 0 kg to 50 kg, using a step of 1 kg.
In column B generate the equivalent value in pounds.
Use your values to convert:
(a) 5 kg to lbs   **(b)** 33 lbs to kg   **(c)** 19 kg to lbs
(d) 4 kg to lbs   **(e)** 101.2 lbs to kg   **(f)** 94.6 lbs to kg

You only need to enter two formulae:
■ =A2+1 in cell A3
■ =A2*2.2 in cell B2

	A	B
1	Kilograms	Pounds
2	0	0
3	1	2.2
4	2	4.4
5	3	6.6

2 To convert kilometres, km, into miles you can divide by 1.6.

In column C generate 0 km to 8 km, using a step of 0.1 km.
In column D generate the equivalent value in miles.
Use your values to convert:
(a) 1 km to miles   **(b)** 1 mile to km   **(c)** 4.5 km to miles
(d) 5 miles to km   **(e)** 6.4 km to miles   **(f)** 4.25 miles to km

3 To convert degrees Celsius, °C, into degrees Fahrenheit, °F, you can multiply by 1.8 and then add 32.

In column E generate 0°C to 30°C, using a step of 1°C .
In column F generate the equivalent value in Fahrenheit.
Use your values to convert:
(a) 10°C to °F   **(b)** 5°C to °F   **(c)** 59°F to °C
(d) 77°F to °C   **(e)** 30°C to °F   **(f)** 48.2°F to °C

Enter 'Celsius' in cell E1, and 'Farenheit' in cell F1.
Enter '0' in cell E2
Enter the formula = e2+1 in cell E3 and drag down to E32.
Enter the formula =e2*1.8+32 in cell F2 and drag down to cell F32.

4 Write down a formula which will change:
(a) lbs to kg   **(b)** miles to km   **(c)** °F to °C

Hint: What is the reverse of multiply by 2.2?

5 Extension: Use the Internet to obtain some currency exchange rates. One suggested site is **http://finance.yahoo.com/m3?u**
(a) Use the rates to create a spreadsheet to change:
■ up to 50 US dollars to Japanese yen
■ up to 50 pounds sterling, £, to euros

(b) Create a line graph showing the conversion from the euro to the £.

Exchange rates can also be found in most daily national newspapers

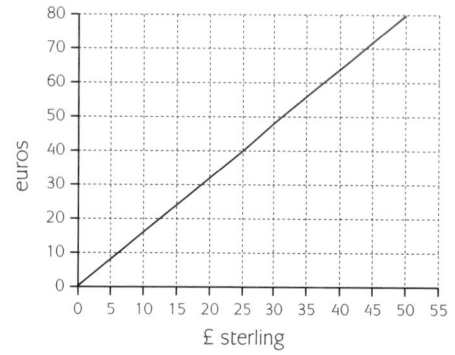

A line graph showing the conversion of pounds sterling to euros

A trail of primes

In this activity sheet you will use the internet to investigate prime numbers.

Click on the internet links below to visit sites that will help you...

http://www.utm.edu/research/primes/largest.html

... find out the number of digits contained in the largest known prime number and who Mersenne was and what his primes are;

http://www.faust.fr.bw.schule.de/mhb/eratosiv.htm

... search for primes using Eratosthenes' sieve;

Mersenne

http://www.math.princeton.edu/~arbooker/nthprime.html

... find the value of the 10 000th prime number and many more;

http://www.utm.edu/research/primes/notes/by_year.html

... find the value of the largest known primes before electronic computers;
... find when we will have a one billion digit prime;

1	2	3	4	5	6	7	8
21	22	23	24	25	26	27	28
41	42	43	44	45	46	47	48
61	62	63	64	65	66	67	68
81	82	83	84	85	86	87	88
101	102	103	104	105	106	107	108
121	122	123	124	125	126	127	128
141	142	143	144	145	146	147	148

http://www.utm.edu/research/primes/programs/music/listen
http://www.2357.a-tu.net/index.php?link=Music

... listen to the music of primes at the above two sites;

http://www.worldofnumbers.com/palpri.htm

... find Palindromic primes;

http://www.worldofnumbers.com/circular.htm

... find Circular primes;

http://primes.utm.edu/glossary/page.php/Permutable
Prime.html

... find Permutable primes.

Activity sheet NA4

Investigating decimal fractions

This activity sheet shows you how to use a calculator and a spreadsheet to work out the decimal equivalent of rational numbers. It includes recurring cycles and digit patterns.

1 To find the decimal equivalent of $\frac{5}{8}$ on your calculator enter:

| 5 | ÷ | 8 | = |

to display 0.625

The decimal equivalent of $\frac{3}{7}$ will be displayed as 0.428571428

This decimal fraction repeats itself every 6 decimal places. It is called a recurring decimal with a cycle of 6.
The decimal fraction continues for ever:

0.428571428571428571428571.......

It is written with a dot above the start and end of the cycle as 0.4̇28571̇

> Any fraction turned into a decimal fraction will either
>
> **terminate like** $\frac{5}{8}$
>
> or
>
> **recur like** $\frac{3}{7}$
>
> The **maximum** recurring cycle a decimal fraction can have is 1 less than the denominator.

2 On your calculator work out the decimal equivalent of $\frac{1}{17}$

Your calculator has not been able to display the full recurring cycle of 0.0588235294117647058....
This decimal fraction has a recurring cycle of 16

Before starting the following activities you need to launch Excel.

3 Open the file decimal.xls and enter the fraction $\frac{1}{17}$ to display the full cycle.

> This spreadsheet can display a decimal fraction up to 20 000 decimal places!

4 $\frac{1}{97}$ has a maximum recurring cycle. What are the last 10 digits in the cycle?

5 With the same denominator, does a change in the numerator affect the size of the cycle?

6 There are 10 other maximum cycles if the denominator is less than 150. Can you find them? [Hint: prime denominators]

7 What is the largest cycle you can find?

> The largest cycle found by a pupil in our school is **39 988**.
> Can you beat it?
> It can be beaten!

(continued)

8 What pattern is there in digits of the recurring cycles of

(a) $\frac{1}{7}, \frac{2}{7}, ..., \frac{6}{7}$?

(b) $\frac{1}{17}, \frac{2}{17}, ..., \frac{16}{17}$?

9 Are there similar patterns for other decimal fractions with maximum recurring cycles?

You can use worksheet SSM20 to find an estimate for the value of π.

10 What patterns are there in the decimal fractions for

$\frac{1}{13}, \frac{2}{13}, ..., \frac{12}{13}$?

Irrational numbers

Some numbers **cannot** be written as decimal fractions because it is impossible to find a fraction which generates them. These numbers are called *irrational* numbers.

π and $\sqrt{2}$ are two famous irrational numbers. Their decimal equivalent can only be approximated.

$\frac{22}{7}$ is a common approximation for π.

You can find more about irrational numbers at

http://mathworld.wolfram.com

Activity sheet NA5

Running an activity weekend

This activity sheet shows you how you can use a spreadsheet to model a situation. Before starting the activity, launch Excel and open the spreadsheet activityweekend.xls

The spreadsheet contains information on a school activity weekend.

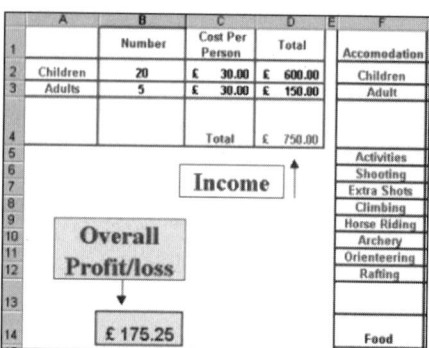

■ The total income, found in cell D4, is from the number of children and adults attending.

■ The total expenditure, found in cell I16, is for food, accommodation and the activities provided.

■ The overall profit or loss, found in cell B14, is found by subtracting the total expenditure from the total income.

At the moment the cost of the weekend is £30 for both children and adults (cells C2 and C3).

1 Since the adults are supervising it is decided they do not need to pay their £30 for the weekend. Amend cell C3 and write down the new total for the income found in cell D4:

£ _____

2 Two children suddenly become ill just before the weekend. Amend cell B2 and increase the cost in cell C2 until the income is as close as possible to the value you entered above. The new total for the income is: £ _____

3 The local supermarket offers a reduced cost for food. Change the value in cell H15 to £2.50. What is the new overall profit/loss found in cell B14?

£ _____

Change the food cost here.

	Number	Cost Per Day	Total
Food	23	£ 5.00	£ 115.00

4 The group decides to try another activity. Use cells G12, H12 and I12 to include the extra 'Rafting' activity shown on the right. You do not need to enter the '£' signs in cells H12 and I12.

The total in cell I16 will change automatically.

Rafting	1	£ 20.00	£ 20.00
		Total	£ 543.75

5 What is the new overall profit/loss found in cell B14?

£ _____

Create your own modelling project

Jim, Aurore and Imke are all thinking of buying a mobile phone. Jim works as a salesman and will be using his phone Monday to Friday during business hours. Aurore will only use her phone for emergencies. Imke's boyfriend lives in another town and she will spend a lot of time chatting with him in the evenings and at weekends. They estimate that they will spend the following amounts of time on their new phones:

Jim	Aurore	Imke
4 hours a day	30 minutes a month	$2\frac{1}{2}$ hours each evening plus 5 extra hours at the weekend

Set up a spreadsheet to model the cheapest way to pay for their new phones and calls for a period of 12 months.

Use the billing information from these websites to help:

http://www.vodafone-retail.co.uk/

http://www.one2one.co.uk

http://www.orange.co.uk

When you buy a phone, you can
- pay for your calls in advance and pay no line rental

or
- pay a monthly line rental and be billed for your calls.

Personal finances and percentages

This activity sheet uses the internet and spreadsheet modelling to help you understand the process of buying a house. Before starting this activity, you should go to the internet site:

http://www.your-move.co.uk/html/default.htm

Stefan and Shamin are thinking of buying a house.

Stefan has an annual salary of £28 500 and Shamin has an annual salary of £32 000. Use the Mortgage calculator link to calculate approximately how much they can borrow.

Enter this value: £_____

They decide to buy a property for about 85% of the amount above.

Enter this value: £_____

Click the   **Property Search**

link and find a suitable house in *your* area or a region near you.

Stefan and Shamin allow £3 300 for miscellaneous expenses and have saved £7 200 for a deposit. Subtract this deposit from the cost of the house and add the miscellaneous expenses to find how much Stefan and Shamin will need to borrow:

Enter this value: £_____

Click ◀ **BACK**

to see what their monthly repayments will be using the *estimate repayments* link.

Choose one of the *repayment* options: £_____

■ Open the Excel file house.xls.

■ Enter all the values from your sheet and all the missing values.

■ Their monthly profit/loss is: £_____

■ The following year Stefan received an 8% pay rise and Shamin received a rise of 6.5%.

■ Monthly expenses, though, went up by 13%.

■ Amend your spreadsheet to find their new monthly profit/loss:

    £_____

| Mortgage Calculator **FOR SALE**

Use your calculator to find 85% by multiplying by 0.85

Miscellaneous expenses can include:
■ solicitor's fees
■ a removal firm
■ estate agent fees

Activity sheet NA8

Money

This activity sheet shows you how to use a spreadsheet for modelling. It includes work on VAT and bar charts. Before starting the activities you need to launch Excel.

Open the spreadsheet file starters.xls which displays a list of starters and their cost available at the New Delhi Tandoori Restaurant. The spreadsheet also shows the number of each type of starter sold in the week commencing 10th July 2000.

1 Enter the formula =B2*C2 into cell D2 to calculate the amount of money received for lentil soup that week.

To multiply numbers in Excel, use the * key.

2 Select cell D2 and drag the black square in the bottom right-hand corner of the cell down to cell D21. This will calculate the amount of money received for the rest of the starters that week.

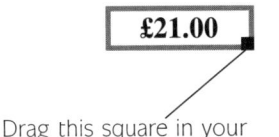

Drag this square in your spreadsheet down to cell D21

3 Enter the formula =Sum(C2:C21) in cell C22 to help calculate the total number of starters sold that week

4 Use a similar technique in cell D22 to calculate the total income for all the starters that week.

5 Ali has to charge his customers VAT on the food sold. This adds 17.5% to the cost. In cell E22 enter the formula =D22*1.175. This will add VAT for you. Enter this value in cell G10.

6 Sales for the next three weeks are shown in the table. For each week change the entries in column C and enter the new totals, found in cell E22, into cells G11, G12 and G13.

7 Highlight the cells G10, G11, G12 and G13 and create a bar chart showing the weekly income from the starters sold. Print out your chart.

	17/7	24/7	31/7
Lentil Soup	12	10	17
Mulligatawny Soup	4	5	10
Prawn Cocktail	9	13	4
Shish Kebeb	12	16	13
Sheek Kebab	12	9	11
Shami Kebab	10	14	14
Lamb Tikka	13	22	21
Chicken Tikka	25	19	19
Tandoori Chicken	39	25	35
Meat Samosa	18	12	17
Vegetable Samosa	10	19	18
Onion Bhajee	31	24	30
Chicken Chatt	11	10	12
Chana Chatt	3	7	8
Prawn Puri	10	6	14
King Prawn Puri	7	8	16
Pakara	12	14	10
Stuffed Pepper	9	9	8
Tandoori Mixed Grill	24	20	17
Vegetable Korma	23	25	10

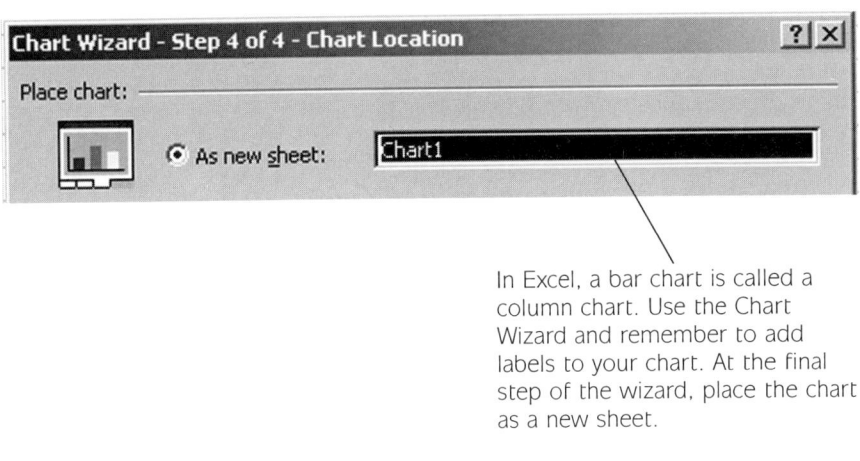

In Excel, a bar chart is called a column chart. Use the Chart Wizard and remember to add labels to your chart. At the final step of the wizard, place the chart as a new sheet.

Activity sheet NA9

Analysing food

This activity sheet asks you to find information from the internet and to use a spreadsheet. It includes work on unit conversions, ratios and proportions.

Go to the internet site:

http://www.taste.co.uk/tastehomepage

and use the recipe search on the recipes link.

Enter 'barbecue baked chicken and beans' in the 'I want to cook' box and click on the go button to bring up the link to the following recipe:

> *Barbecue baked chicken and beans*
> *Serves: 4*
> *Make and cook time: approximately 30 minutes*
>
> 4×160 g chicken drumsticks
> 4×75 g chicken thighs
> 1×850 g can baked beans
> 1 medium onion
> 1 garlic clove
> 1 tbsp vegetable oil
> 150 ml tomato ketchup
> 3 tbsp American mustard
> 2 tbsp Worcestershire sauce
> 1 tbsp treacle
> salt and freshly ground black pepper

1 Use the *Conversion table* link to change the recipe above by converting:

grams, g $\longrightarrow$ ounces, oz
and millilitres, ml $\longrightarrow$ fluid ounces, fl oz

2 Find another recipe for 4 people. Print out a copy and change the ingredients so that the meal is suitable for

(a) 8 people   **(b)** 20 people   **(c)** 10 people   **(d)** 6 people

You could copy the ingredients and paste them into your word processing package and make the necessary changes there.

Before starting the following activities you need to launch Excel software.

3 Use a spreadsheet to produce five bar charts comparing the energy produced and each of the four nutrient types in the table on the following page. Use the ctrl key to highlight columns which are not next to each other.

(continued)

Ingredient (100g):	Energy produced in kJ	Fat in g	Protein in g	Calcium in mg	Carbohydrate in g
1 medium banana	403	0.3	1.2	6.0	23.2
100g butter	3031	81.7	0.5	15.0	0
2 small chapatis	1383	12.8	8.1	66.0	48.3
$2\frac{1}{2}$ slices chicken	599	4.0	26.5	9.0	0
16 tbsp cornflakes	1535	0.7	7.9	15.0	85.9
10 squares milk chocolate	2214	30.3	8.4	220.0	39.4
3 slices wholemeal bread	914	2.5	9.2	54.0	41.6
$\frac{1}{2}$ glass semi-skimmed milk	195	1.6	3.3	120.0	5.0

4 Joshua woke up late and was very hungry. For his breakfast he ate
- 2 medium bananas
- 24 tablespoons of cornflakes
- 2 chapatis
- 3 slices of wholemeal bread
- 1 g of butter
- 5 slices of chicken breast
- the equivalent of $1\frac{1}{2}$ glasses of milk

Use your spreadsheet to calculate the amount of energy produced by the food and the total of each type of nutrient that Joshua had for his breakfast.

Use the Ctrl key on your keyboard to help you highlight columns that are not next to each other.

Fat is a source of energy that your body stores easily.
Protein is essential for the building of new body cells.
Calcium is good for the bones.
Carbohydrates convert to energy very quickly.

Activity sheet NA10

Investigating quadratics 1

This activity sheet makes use of a graphing package to show you how the value of the integers a, b and c affect the curve $y = ax^2 + bx + c$. Before starting the activities, launch Omnigraph.

1 How does the shape of the quadratic curve
$y = ax^2 + bx + c$ change if:
   (a) $a > 0$?
   (b) $a < 0$?

In these questions you should choose your own values for a, b and c when entering the equation $y = ax^2 + bx + c$

2 Where does the curve $y = ax^2 + bx + c$ cut the y-axis if:
   (a) $c > 0$?
   (b) $c < 0$?

3 Describe the relationship between the curve and the x-axis if:
   (a) $b^2 - 4ac > 0$?
   (b) $b^2 - 4ac = 0$?
   (c) $b^2 - 4ac < 0$?

$b^2 - 4ac$ is called the discriminant of the equation $y = ax^2 + bx + c$

Clear the screen before each part of the following question.

---

4 State the sign of a, c and $b^2 - 4ac$ for each of the curves below. Choose suitable values for a, b and c and find an equation which could produce each curve.

(a) 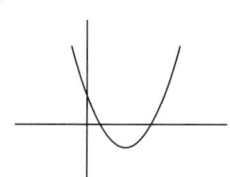   **(b)** 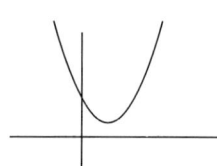   **(c)**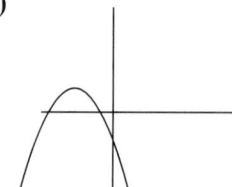

(d) 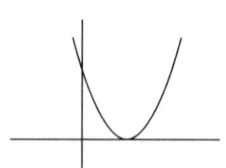   **(e)** 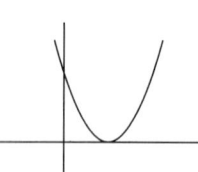   **(f)**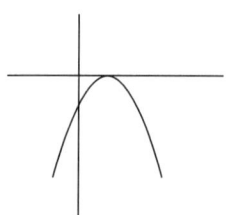

5 Determine whether a quadratic curve could be produced from the information below. Sketch any possible curve on *paper*.
   (a) $a < 0$, $b^2 - 4ac = 0$ and $c < 0$
   (b) $a < 0$, $b^2 - 4ac = 0$ and $c > 0$
   (c) $a > 0$, $b^2 - 4ac > 0$ and $c > 0$
   (d) $a > 0$, $b^2 - 4ac > 0$ and $c < 0$
   (e) $a > 0$, $b^2 - 4ac < 0$ and $c < 0$
   (f) $a > 0$, $b^2 - 4ac < 0$ and $c > 0$

Activity sheet NA11

Investigating sequences

This activity sheet shows you how to use a spreadsheet to generate the Fibonacci sequence and the Golden Number. Before starting the activities, you need to launch Excel.

	A The Fibonacci sequence	B The ratio of successive terms
1		
2	1	1.000000000000
3	1	2.000000000000
4	2	1.500000000000
5	3	1.666666666667
6	5	1.600000000000
7	8	1.625000000000
8	13	1.615384615385
9	21	1.619047619048
10	34	1.617647058824

1 In cells A2 to A26 generate the first 25 terms of the Fibonacci sequence:

1, 1, 2, 3, 5, 8, 13, ...

by entering 1 in cell A2, 1 in cell A3 and the formula = A3+A2 in cell A4. Copy the formula by dragging down to cell A26.

2 In cells B2 to B25 generate the ratio of successive terms

$$\frac{1}{1}, \frac{2}{1}, \frac{3}{2}, \frac{5}{3}, \frac{8}{5}, \frac{13}{8}, ...$$

by entering the formula =A3/A2 in cell B2 and copying the formula down to cell B25.

3 Highlight the cells in column B and create a scatter graph of the ratios

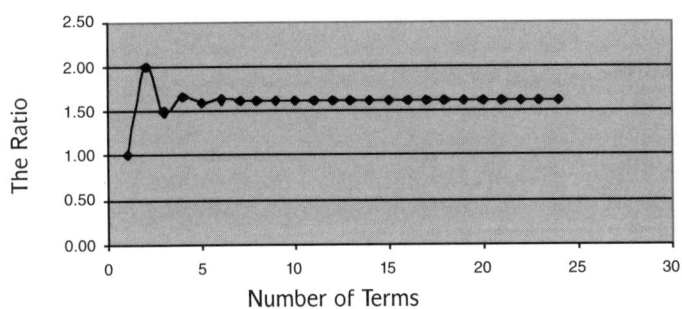

The ratio of successive terms

Widen column B and Format the Cells Number to 11 Decimal places.

4 In Cell C25 enter the value $\frac{\sqrt{5}+1}{2}$ as = (sqrt(5)+1)/2

This value is known as the **Golden Number** and it occurs naturally in many areas of Nature. It was known to the Greeks and they used the *Golden Proportion* in their building designs.

Notice how close its value is to the number in cell B25.

5 Extend the Fibonacci sequence in column A to add another 25 terms. Highlight *any* 10 consecutive terms in the sequence, sum them and divide the total by the *seventh* term of your 10. Try this again with *two* other lots of 10 consecutive terms. What do you notice about your answers?

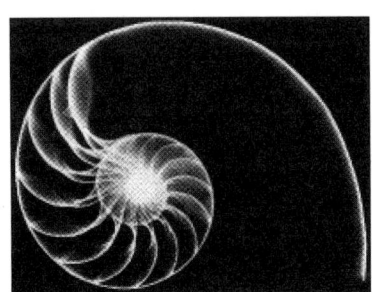

The spiral in this Nautilus shell is a Fibonacci spiral.

6 Find out more about the Golden Number, Golden Rectangle, Fibonacci sequence, Rabbits and Nature... on the internet at

http://www.mcs.surrey.ac.uk/personal/R.Knott/Fibonacci/fib.html

Activity sheet NA12

Linear functions and their inverse

This activity sheet makes use of a graphing package to help you find the inverse of linear functions. Before starting the activities, launch Omnigraph.

A *linear* function is of the type $y = mx + c$, where:
- m is the gradient
- c is the value where the *straight line* $y = mx + c$ crosses the y-axis.

Example: $y = 3x - 2$

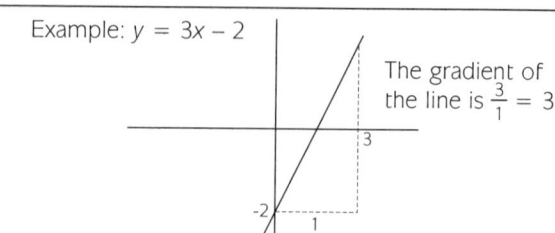

The gradient of the line is $\frac{3}{1} = 3$

The *inverse* of a function is the *reflection* of the function in the line $y = x$.

Click this button to clear the lines as necessary:

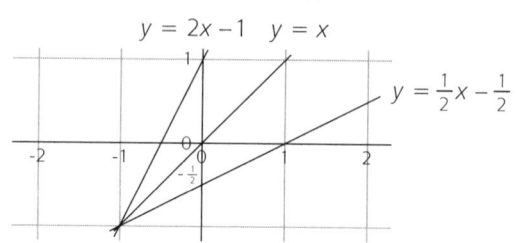

$y = 2x - 1$ $y = x$

$y = \frac{1}{2}x - \frac{1}{2}$

1 Use Omnigraph to help you complete the table below.

	Linear function	Inverse linear function
(a)	$y = 2x$	
(b)	$y = 2x - 1$	
(c)	$y = 2x + 4$	
(d)	$y = \frac{1}{2}x + 1$	
(e)	$y = 3x - 1$	
(f)	$y = 3x + 2$	
(g)	$y = -2x$	
(h)	$y = -2x + 3$	
(i)	$y = 3 - x$	
(j)	$y = 4 - x$	

To find the inverse click on:
- Transform
- Reflect
- Type $y = x$ in the space provided and press Enter
- Work out the gradient using the squares on screen
- Read off the value where the reflected line crosses the y-axis
- Write the inverse function in the table

Hint:
Lines with a *negative* gradient slope this way:

Why do you think the functions in parts (i) and (j) are called *self-inverse* functions?

2 Write down another two self-inverse functions and check your answers on screen.

$y =$ _____ and $y =$ _____

3 Write down the inverse function of $y = mx + c$.

$y =$ _____

Investigating quadratics 2

This activity sheet shows you how to produce quadratic curves by completing the square and using translations. Before starting the activities, you need to launch Omnigraph.

Enter the curve with equation $y = x^2$
The curve turns at $(0, 0)$ and the y-axis is its line of symmetry.

1 Enter the curve with equation
$$y = (x - 1)^2 \quad [\equiv y = x^2 - 2x + 1]$$
   (a) What is the turning point of this curve?
   (b) Describe the transformation of this curve from the curve $y = x^2$.

> These curves are called *quadratic curves* and another name for their turning point is the *minimum* point of the curve.

2 Enter the curve with equation
$$y = (x + 2)^2 \quad [\equiv y = x^2 + 4x + 4]$$
   (a) What is the turning point of this curve?
   (b) Describe the transformation of this curve from the curve $y = x^2$.

3 What is the turning point of these five curves?
   (a) $y = (x + 5)^2$ **(b)** $y = (x - 3)^2$
   (c) $y = (x - 4)^2$ **(d)** $y = x^2 - 4x + 4$
   (e) $y = x^2 + 6x + 9$

> $y = (x - a)^2 \equiv y = x^2 - 2ax + a^2$
> and
> $y = (x + a)^2 \equiv y = x^2 + 2ax + a^2$

4 Enter the equation of the quadratic curves with a minimum point at
   (a) $(-4, 0)$ **(b)** $(5, 0)$ **(c)** $(-3, 0)$ **(d)** $(1.5, 0)$ **(e)** $(-2.5, 0)$

> Clear the screen before each part of the question.

5 Enter the curve with equation $y = x^2$ and the curve with equation $y = (x - 3)^2 + 2$.
   Write down the coordinates of the minimum point of the curve $y = (x - 3)^2 + 2$ and describe the two transformations from the curve $y = x^2$.

> Check your answers on screen.
> $y = (x - 3)^2 + 2$
> $\quad = x^2 - 6x + 9 + 2$
> $\quad = x^2 - 6x + 11$

6 Work out:
   (a) The minimum point of the curve $y = (x + 3)^2 - 2$
   (b) The equation of this curve in the form $y = x^2 + bx + c$

7 Enter the equation of the quadratic curves with a minimum point at:
   (a) $(4, 2)$ **(b)** $(-5, 2)$ **(c)** $(4, -2)$ **(d)** $(-4, -3)$ **(e)** $(-6, 5)$

8 Enter the equations of these curves and use their minimum point to help you complete the square on each equation.
   (a) $y = x^2 - 2x - 2$ **(b)** $y = x^2 + 2x + 7$
   (c) $y = x^2 + 6x + 4$ **(d)** $y = x^2 - 4x - 2$
   (e) $y = x^2 - 10x - 21$

Approximating a value for π

This activity sheet uses a spreadsheet and the internet to investigate series, sequences, convergence, percentage errors and other types of numbers. Before starting the activities, launch Excel.

Use a spreadsheet to find an approximate value for π by summing the first 2000 terms in the following infinite series:

π is an *irrational* number and it is therefore not possible to find an exact value for it.

$$\pi = 4(1 - \frac{1}{3} + \frac{1}{5} - \frac{1}{7} + \frac{1}{9} - ...)$$

In this series the denominators are the sequence of odd numbers.

Enter these formulae in row 3 and copy them down to row 2000:

	A	B	C	D
1	**Numerator**	**Denominator**	**Decimal fraction**	**Cumulative sum**
2	1	1	4	4
3	=A2*-1	=B2+2	=4*A3/B3	=(C3+D2)
4	=A3*-1	=B3+2	=4*A4/B4	=(C4+D3)
5	=A4*-1	=B4+2	=4*A5/B5	=(C5+D4)
6	=A5*-1	=B5+2	=4*A6/B6	=(C6+D5)
7	=A6*-1	=B6+2	=4*A7/B7	=(C7+D6)

Highlight cells A3 – D3 and click and drag the black square in the bottom right of cell D3 to copy the formulae.

The first 7 rows should look like this:

	A	B	C	D
1	**Numerator**	**Denominator**	**Decimal fraction**	**Cumulative sum**
2	1	1	4	4
3	–1	3	–1.333333333	2.666666667
4	1	5	0.8	3.466666667
5	–1	7	–0.571428571	2.895238095
6	1	9	0.444444444	3.33968254
7	–1	11	–0.363636364	2.976046176

After 2000 terms the cumulative sum is:

2001	–1	3999	–0.00100025	**3.141092654**

(continued)

Use a spreadsheet to estimate a value for π in the following infinite series:

It is a good idea to plan your spreadsheet on paper first. Use 2000 terms for each estimate.

1 $\pi = \dfrac{4}{\sqrt{2}}(1 + \dfrac{1}{3} - \dfrac{1}{5} - \dfrac{1}{7} + \dfrac{1}{9} + \dfrac{1}{11} - \dfrac{1}{13} - \dfrac{1}{15} + ...)$

2 $\pi = 2(\dfrac{2}{1} \times \dfrac{2}{3} \times \dfrac{4}{3} \times \dfrac{4}{5} \times \dfrac{6}{5} \times \dfrac{6}{7} \times ...)$

3 $\pi = \sqrt{6\left(1 + \dfrac{1}{2^2} + \dfrac{1}{3^2} + \dfrac{1}{4^2} + ...\right)}$

The series in Question 3 is a rearrangement of the *Basel Problem*: to find the sum of the infinite series where the denominators are the square numbers:

$1 + \dfrac{1}{4} + \dfrac{1}{9} + \dfrac{1}{16} + ...$

The problem was solved by Euler in 1735 when he proved

that the sum was $\dfrac{\pi^2}{6}$. For more on

this problem see:

Euler, The Master of Us All by William Dunham, Chapter 3, published by The Mathematical Association of America.

4 For each approximation calculate the percentage error using the π button on your calculator as the 'true' value of π.

5 Investigate the sum of the infinite series where the denominators are the triangular numbers:

$$1 + \dfrac{1}{3} + \dfrac{1}{6} + \dfrac{1}{10} + ...$$

The triangular numbers can be generated from

$\dfrac{k(k+1)}{2}$, for $k = 1, 2, 3, ...$

6 Use the internet site:

http://mathworld.wolfram.com

to find out:
- more about π by beginning with a search for pi
- more about irrational numbers
- what *transcendental* numbers are
- what *imaginary* numbers are
- what the transcendental number *e* is
- the connection between *e*, *i* and π, namely Euler's 'beautiful' formula, also found in *Euler, The Master of Us All*

Infinite series

This activity sheet shows you how to use a spreadsheet to investigate the divergence and convergence of series. Before starting the activities you need to launch Excel.

The infinite series below is called the *harmonic series:*

$$1+\frac{1}{2}+\frac{1}{3}+\frac{1}{4}+\frac{1}{5}+...$$

In this series the denominators are the sequence of *natural numbers*.

Open a new spreadsheet and enter the following in cells A1 to C3:

	A	B	C
1	**Denominator**	**Decimal fraction**	**Cumulative sum**
2	1	=1/A2	1
3	=A2+1	=1/A3	=C2+B3

■ highlight cells A3, B3 and C3 and copy down to row 21
■ highlight columns B and C and click Format and then Cells and change the Number of Decimal places to 15.

1 Write down the approximate sum of the first 20 terms.

2 What is the approximate sum of the first 200 terms?

3 Create a line graph displaying the cumulative sums up to 200 terms and comment on whether or not you think the series is diverging or converging.

4 What is the approximate sum of the first 2000 terms?

5 After how many terms does the sum pass 10?

6 After how many terms does the sum pass 11?

7 Investigate the limit of the alternating series $1-\frac{1}{2}+\frac{1}{3}-\frac{1}{4}+\frac{1}{5}-...$ by using the formulae displayed below.

	A	B	C	D
1	**Denominator**	**Numerator**	**Decimal fraction**	**Cumulative sum**
2	1	1	=B2/A2	1
3	=A2+1	=B2*-1	=B3/A3	=C3+D2

A *diverging* series has no limit as the number of terms increases. $1 + 3 + 5 + 7 + 9 + ...$ is a diverging series.

A *converging* series has a definite finite limit no matter how many terms are added

$1 + \frac{1}{2^2} + \frac{1}{3^2} + \frac{1}{4^2} + ...$ converges to $\frac{\pi^2}{6}$

(continued)

8 Create a line graph displaying the sum of the first 200 terms.

9 Investigate the limit of the series $1 + \frac{1}{2} + \frac{1}{4} + \frac{1}{8} + \frac{1}{16} + ...$

10 Investigate the series $1 + \frac{1}{2^k} + \frac{1}{3^k} + \frac{1}{4^k} + \frac{1}{5^k} + ...$
for:
(a) $k > 1$
(b) $0 < k < 1$
(c) $k < 0$
(d) $k = 1$
Comment on your findings.

Hints for Question 10:
■ Use the formulae shown.
■ Choose an initial value for k which is greater than 1.
■ The $ sign stops D3 becoming D4, D5, ... in column B.
■ Create a line graph for column C.
■ Enter different values for k and see what happens.

	A	B	C	D
1		**Denominator**	**Cumulative sum**	
2	1	=1/(A2^D3)	=B2	k
3	=A2+1	=1/(A3^D3)	=B3+C2	2
4	=A3+1	=1/(A4^D3)	=B4+C3	

The harmonic series is very slow to diverge. It is *crawling* towards infinity!

■ After 2.5×10^8 terms the sum is still less than 20

■ To reach a sum greater than 100 the number of terms needed is 15,092,688,622,113,788,323,693,563,264,538,101,449,859,497, a number to be found in *The Sixth Book of Mathematical Games from Scientific American*, by Martin Gardner (1984) Chicago, IL: University of Chicago Press, page 167.

Activity sheet NA16

Solving equations

In this activity sheet you will be working with cubic equations and straight lines. Before starting the activities, launch Omnigraph.

To enter the cubed sign [or the squared sign] in Omnigraph, hold down the Alt key on your keyboard as you press 3 [or 2]. Alternatively you can type $x \wedge 3$ [or $x \wedge 2$].
Or click this button to add powers.

1 Enter the curve with equation $y = x^3 - 4x^2 + x + 6$

   (a) Use the curve to solve the equation
$$x^3 - 4x^2 + x + 6 = 0$$

   (b) Enter the line $y = 2$ and use it to find an approximate solution to the equation
$$x^3 - 4x^2 + x + 6 = 2$$

2 Enter the curve with equation
$$y = 2x^3 + 5x^2 + x - 2$$

Remember the box below the axes is where you enter equations.

   (a) Use the curve to solve the equation
$$2x^3 + 5x^2 + x - 2 = 0$$

   (b) Add a suitable line to help find an approximate solution to the equation
$$2x^3 + 5x^2 + x - 2 = -3$$

Use your mouse to move the cursor over the intersection points and read off the co-ordinates on the bottom menu bar.

3 Enter the curve with equation
$$y = x^3 - 6x^2 + 9x - 4$$

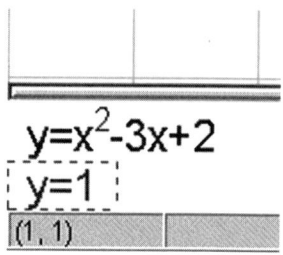

   (a) Use the curve to solve the equation
$$x^3 - 6x^2 + 9x - 4 = 0$$
   and explain why there are only two solutions.

   (b) Add the line $y = x + 2$ to help find the approximate solution to the equation
$$x^3 - 6x^2 + 9x - 4 = x + 2$$

   (c) Hence solve the equation $x^3 - 6x^2 + 8x - 6 = 0$.

Hint for Question **3(c)**:
Rearrange the equation in **(b)** to leave zero on the right-hand side.

4 Solve the following equations on screen using *only* the curve $y = x^3 + 2x^2 - x - 2$ and a straight line when necessary.

   (a) $x^3 + 2x^2 - x - 2 = 0$

   (b) $x^3 + 2x^2 - x - 2 = x - 1$

   (c) $x^3 + 2x^2 - 2x - 1 = 0$

   (d) $x^3 + 2x^2 - 3x - 3 = 0$

The approximate solution to equations

This activity sheet shows you how to use a spreadsheet to solve equations by iteration. Before starting the activities you need to launch Excel.

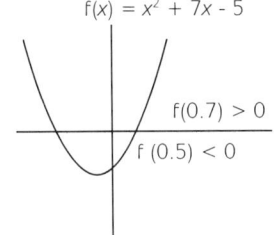

$f(x) = x^2 + 7x - 5$

$f(0.7) > 0$

$f(0.5) < 0$

The quadratic equation $x^2 + 7x - 5 = 0$ has a solution *close* to $x = 0.6$. This means that the curve crosses the x-axis close to this value of x.

If $f(x) = x^2 + 7x - 5$ then $f(0.7) > 0$ and $f(0.5) < 0$. The value of x where $f(x) = 0$ is therefore between 0.5 and 0.7.

You can use a spreadsheet to find an approximate solution to the equation $x^2 + 7x - 5 = 0$

Rearrange the equation as follows:

$x^2 + 7x = 5$

$7x = 5 - x^2$

$x = \dfrac{5 - x^2}{7}$

Set up a spreadsheet using the formulae below:

	A	B
1	x	$(5 - x^2)/7$
2	0.5	=(5-A2^2)/7
3	=B2	=(5-A3^2)/7

To key in x^2, type x then hold down the Alt key on your keyboard as you type 253 on the number pad. *Make sure the Num Lock is on.*

■ Enter 0.5 in cell A2 as your starting value.

■ Cell B2 calculates $\dfrac{5 - 0.5^2}{7} \approx 0.678571428$

■ This improved value for your approximate solution to the equation is then transferred to cell A3.

■ Cell B3 then calculates $\dfrac{5 - 0.678571^2}{7} \approx 0.64850583$, an even better approximate solution to the equation.

■ Copying down row 3 to row 16 continues this process where the values in cells A16 and B16 are equal to 9 decimal places.

■ This gives an approximate solution to the quadratic equation $x^2 + 7x - 5 = 0$ of $x = 0.653311931$.

Highlight columns A and B and select Format, Cells. Use the Number tab Category: Number and increase the Decimal places to 9. Widen the columns if necessary.

Make sure columns A and B are wide enough to show 9 decimal places.
Remember you can copy a formula by dragging.

(continued)

Use a spreadsheet to find an approximate solution to the following equations:

Give your answer correct to 9 decimal places.

Question	Equation	Rearrangement	Starting value
1	$x^2 - 9x + 6 = 0$	$x = \frac{x^2 + 6}{9}$	1
2	$x^2 - 7x + 2 = 0$	$x = \sqrt{7x - 2}$	7
3	$x^3 + 2x^2 + 5x - 1 = 0$	$x = -\frac{(x^3 + 2x^2 - 1)}{5}$	0
4	$x^3 + 3x - 15 = 0$	$x = \sqrt[3]{15 - 3x}$	2
5	$-x^4 + 4x + 3 = 0$	$x = \sqrt[4]{4x + 3}$	1.7

Hints:

In Question 2 use the formula
`=SQRT(7*A2-2)` in the Rearrangement column and continue as in the example on p119.

In Question 3 use the formula
`=-(A2^3+2*A2^2-1)/5` in Rearrangement column.

In Question 4 use the formula
`=(15-3*A2)^(1/3)` in Rearrangement column.

Activity sheet NA18

Odd and even functions

This activity sheet makes use of a graphing package to investigate odd and even functions. Before starting the activities, launch Omnigraph.

■ An *even* function has the *y*-axis as a line of symmetry.

■ $f(x)$ is even if $f(x) = f(-x)$

■ An *odd* function has rotational symmetry following a rotation of 180° about the origin.

■ $f(x)$ is odd if $f(-x) = -f(x)$

1 Investigate when the function $y = x^n$ is:
    (a) odd   **(b)** even
       for $n = 1, 2, 3, ...$

2 Use your answers to Question 1 to decide whether the following are odd or even functions.

    (a) The product of two even functions is _____

    (b) The difference of two even functions is _____

    (c) The product of two odd functions is _____

    (d) The product of an odd and an even function is _____

Remember:
$x^n \times x^m = x^{n+m}$

Click this button to clear the functions as necessary:

3 Complete the table below.

Function	odd or even
$y = \sin x$	
$y = \cos x$	
$y = \sin x \cos x$	
$y = \dfrac{1}{x^2}$	
$y = \dfrac{1}{x}$	
$y = x - \dfrac{1}{x}$	
$y = x^2 + \dfrac{1}{x^2}$	
$y = x^3 + \dfrac{1}{x^3}$	
$y = x^2 + \dfrac{1}{x^3}$	

When entering *trigonometric* functions:

■ Change the axes to trigonometric by clicking Zoom and then Trig scales.

To change back:

■ Click Zoom and then Reset scales.

Click this button to add powers

4 Create some odd and even functions of your own and check them on screen.

Function transformations

This activity sheet makes use of a graphing package to investigate function transformations. Before starting the activities, launch Omnigraph.

> Change the axes to trigonometric by clicking Zoom, then Trig scales. Zoom and Rescale the x-axis from –720° to 720° and the y-axis from –5 to 5.

1 Enter the curve $y = \sin x$.

In the table below, describe the single transformation or combined transformations from the curve $y = \sin x$ to the new curve.

Choose from the following transformations and state any scale factors used:

■ a stretch parallel to the y-axis
■ a stretch parallel to the x-axis
■ a translation parallel to the y-axis
■ a translation parallel to the x-axis
■ a reflection in the x-axis
■ a reflection in the y-axis.

New curve	Transformation(s)
$y = \sin 2x$	
$y = 2\sin x$	
$y = \sin \frac{1}{2}x$	
$y = \frac{1}{2}\sin x$	
$y = \sin x + 3$	
$y = \sin x - 4$	
$y = 3\sin \frac{1}{2}x$	
$y = \frac{1}{3}\sin 3x - 2$	

Click this button to clear the functions as necessary:

Click this button to add a fraction

Click this button to add powers

Clear all curves, click on Zoom and Reset scales.

(continued)

2 Fill in the white cells in the table below to describe the single transformation or combined transformations from the curves $y = 2^x$ and $y = x^3$ to the new curve.

New curve	Transformation(s) from $y = 2^x$	Transformation(s) from $y = x^3$
$y = 2^{-x}$		
$y = -2^x$		
$y = (x - 3)^3$		
$y = (x + 2)^3$		
$y = 2^x + 1$		
$y = (x - 2)^3 + 1$		
$y = (x + 2)^3 - 1$		
$y = 2^{2x}$		

3 If $a > 0$ and $b > 0$, describe the transformation from $y = f(x)$ to:
(a) $y = f(x + a)$
(b) $y = f(x - b)$
(c) $y = f(x) + a$
(d) $y = f(x) - b$
(e) $y = af(x)$
(f) $y = f(ax)$

4 Describe the transformation from $y = f(x)$ to:
(a) $y = -f(x)$
(b) $y = f(-x)$

Activity sheet SSM 1

Plotting coordinates 1

This activity sheet is about creating polygons. Before starting the activities launch Omnigraph.

1 Choose appropriate coordinates to create the following quadrilaterals.

Question	Quadrilateral
(a)	A rectangle
(b)	A trapezium
(c)	An irregular convex quadrilateral
(d)	An irregular concave quadrilateral
(e)	A kite
(f)	A parallelogram

A *convex* polygon has no interior angles more than 180°
The shape below is a convex polygon:

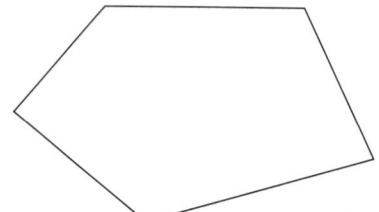

2 Choose appropriate coordinates to create the following triangles.

Question	Triangle
(a)	A scalene triangle
(b)	An isosceles triangle
(c)	A right-angled scalene triangle
(d)	A right-angled isosceles triangle

A *concave* polygon can have an interior angle more than 180°
The shape below is a concave polygon:

3 Choose appropriate coordinates to create the following polygons.

Question	Polygon
(a)	A concave octagon
(b)	A concave heptagon
(c)	A decagon
(d)	A concave nonagon
(e)	A dodecagon

A dodecagon is a polygon with 12 sides.

Activity sheet SSM2

Plotting coordinates 2

This activity sheet is about creating polygons and lines of symmetry.
Before starting the activities launch Omnigraph.

To draw lines in Omnigraph,
select:
Shapes
and then
the appropriate polygon.

1 Create the following polygons.

Question	Polygon	Side length	Centre
(a)	Triangle	2	(5, 0)
(b)	Square	2	(2, 2)
(c)	Pentagon	2	(–4, 0)
(d)	Hexagon	2	(2, –3)
(e)	Octagon	1	(–4, –5)

Use the hand button to move the
grid as required.

2 Start a new Window and recreate the polygons below on screen
by choosing the necessary coordinates.
Select Shapes, New shape to bring up the coordinate editor.

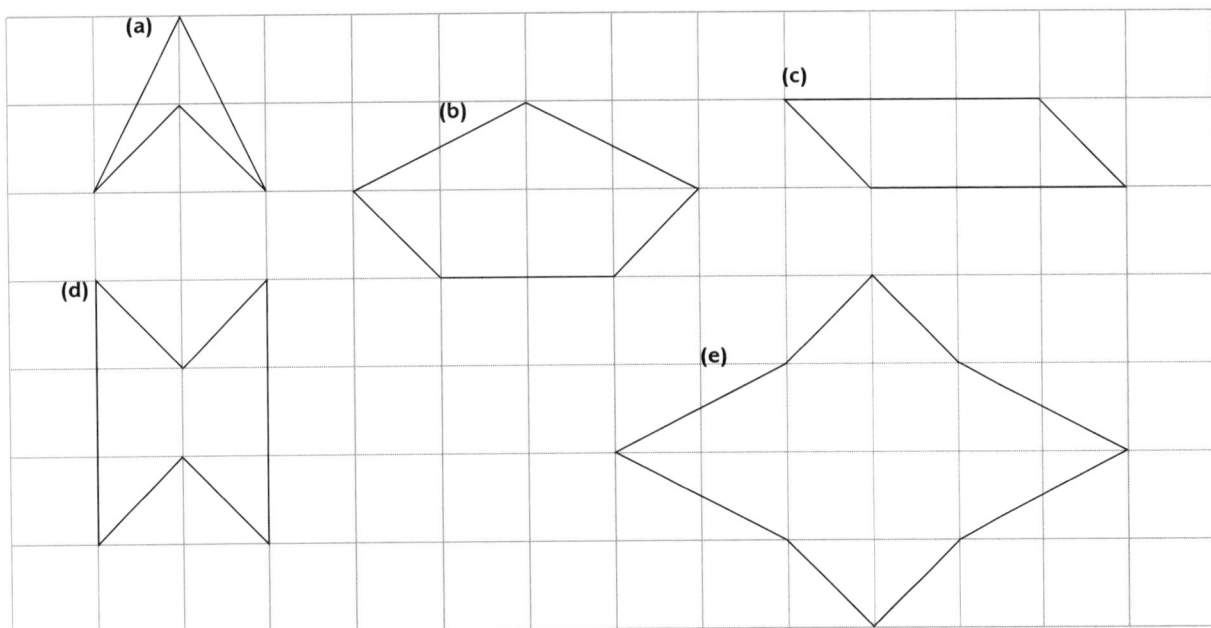

3 Add any lines of symmetry to the polygons. Click on Custom then
Comment box and add the mathematical name of each polygon,
stating whether they are concave or convex.

Horizontal and vertical lines can
be drawn by selecting Analysis and
then Straight lines

4 Print out your shapes.

Investigating angles 1

This activity sheet investigates angles on a straight line and angles round a point.

Launch The Geometer's Sketchpad and:

- Draw a horizontal line AB.
- Add a point C to the line.
- Add a second point D above the line.
- Select the points C and D and Construct a Segment to produce the line CD.
- Measure the size of the angles, ACD and BCD.
- Use the calculator to display the sum of the two angles by ringing up the calculator and clicking on m∠ACD =135° then + button then m∠BCD =45°.

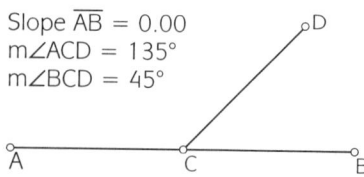

Slope $\overline{AB}$ = 0.00
m∠ACD = 135°
m∠BCD = 45°

To measure angles, the arrow in the menu bar should be selected. Points need to be selected in the correct order.

1 What do you notice about the sum of the two angles as you drag the point D round the screen?

2 Is the sum of the two angles still the same if you change the slope of the line AB by dragging the point B round the screen?

Draw a pair of intersecting lines and add a point where they intersect as shown:

Measure the size of the *vertically opposite* angles, FJH and IJG.

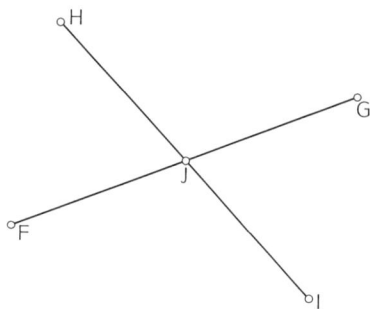

3 What do you notice about their size?

4 Is the same property true for the vertically opposite angles FJI and GJH?

5 Is the same property still true if you rotate either of the two lines by dragging any of the points F, G, H or I round the screen?

Open a New Sketch.
Draw a line AB and add a point C close to the centre.
Add a point D below the line AB.
Select the points C and D and Construct a Segment to produce the line CD.
In a similar way construct the lines CE and CF.

m∠ACD = 41°
m∠DCF = 52°
m∠FCB = 87°
m∠BCE = 95°
m∠ECA = 85°

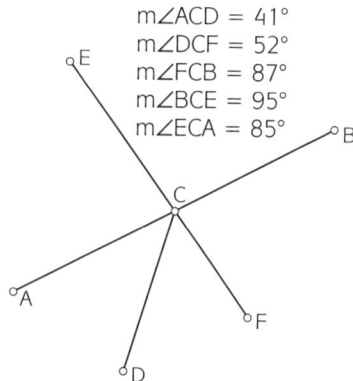

6 Use the calculator to find the sum of the angles ACD, DCF, FCB, BCE and ECA.

7 Investigate the sum of the 5 angles as you drag the point B round the screen.

Do not let any of the lines cross each other as you drag the point B.

Activity sheet SSM 4

Estimating angles

This activity sheet asks you to estimate and check the size of acute, obtuse and reflex angles. Before starting the activities launch The Geometer's Sketchpad.

1 Draw five different *acute* angles. Write down an estimate of their value.
Use the program to <u>M</u>easure your angles and check your estimates.

2 Draw five different *obtuse* angles. Write down an estimate of their value.
Use the program to <u>M</u>easure your angles and check your estimates.

A reflex angle has a value of more than 180° but Geometer's Sketchpad only measures angles less than 180°.

3 Draw and label five different *reflex* angles. Write down an estimate of their value.
Use the program to <u>M</u>easure your angles – it will measure the acute or obtuse part. *Subtract* these measurements from 360° to check your estimates.

Use the Calculator to display the subtraction.

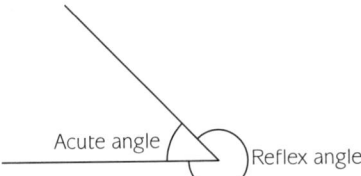

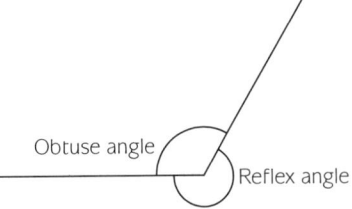

Activity sheet SSM 5

Drawing angles and lines

In this activity sheet you will draw acute, obtuse and reflex angles, parallel and perpendicular lines, equilateral triangles, squares and parallelograms. Before starting the activities, launch MSWLogo.

1 Draw:
    (a) an *obtuse* angle of 120°
    (b) an *acute* angle of 40°
    (c) a *reflex* angle of 320°

2 Draw:
    (a) a pair of *parallel* lines
    (b) a pair of *perpendicular* lines
    Use the penup (**pu**) and pendown (**pd**) commands, to help.

3 Draw:
    (a) an *equilateral triangle* of side 100 units
    (b) a *square* of side 120 units
    (c) the *parallelogram* shown below:

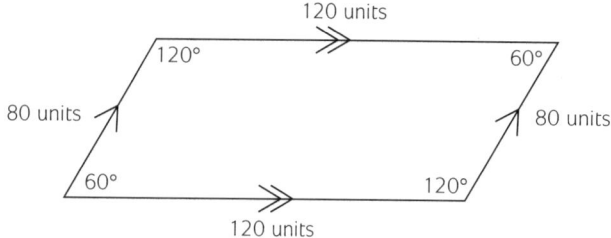

4 Draw the initials of your name, for example:

It is a good idea to sketch your angles on paper first:
To draw an acute angle of 50°, type
    rt 90
    fd 100
    lt 130
    fd 100

Make the arms of your drawings a length of 100 units

In question 4 make use of:
penup or pu to move between letters and
pendown or pd before drawing your next letter.

Activity sheet SSM6

Investigating angles 2

This activity sheet investigates corresponding angles, alternate angles and parallelograms. Before starting the activities launch The Geometer's Sketchpad.

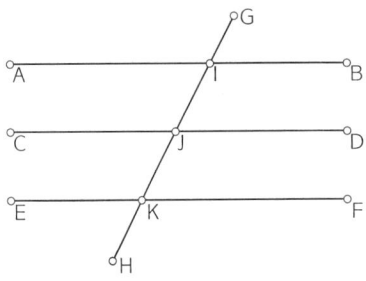

Draw three horizontal lines AB, CD and EF.
Draw a diagonal line GH, which cuts across all the other lines.
Add points I, J and K where the line GH intercepts the three horizontal lines.
Measure the size of the corresponding angles GIB, IJD and JKF.

To measure angles, the arrow in the menu bar should be selected. Points need to be selected in the correct order.

1 What do you notice about the size of the three *corresponding* angles?

2 Rotate the line GH and note the sizes of the three angles. Is the same property true?

3 Measure the size of the pair of alternate angles AIJ, IJD. What do you notice about their size?

4 Measure the size of another pair of alternate angles. Is the same property true?

Horizontal lines have a gradient [slope] of zero

5 What do you notice about the sum of the pair of *interior* angles BIJ and IJD? The angles are *supplementary*.

Slope $\overline{AB}$ = 0.00
Slope $\overline{CD}$ = 0.00
Slope $\overline{EF}$ = 0.00

6 Check the sum of another pair of interior angles.

On a new sketch, draw a horizontal line AB.
Copy the line and paste a second identical line.
Join the two lines to make a parallelogram ABCD.
Measure the size of the four interior angles, BAC, ACD, CDB and DBA, of your parallelogram.

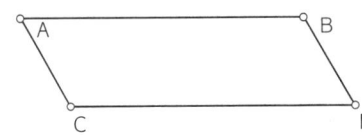

7 Explain why the sum of these four angles is 360°.

8 Use the calculator to display the sum of the four interior angles.
Select the vertices A **and** B and use the mouse to drag the point A to change the size of the parallelogram. Does the sum remain the same?

m∠BAC + m∠ACD + m∠CDB + m∠DBA = 360°

Activity sheet SSM7

Investigating angles in triangles and quadrilaterals

This activity sheet investigates the angles in triangles and quadrilaterals. Before starting, launch Geometer's Sketchpad.

Draw a triangle ABC and <u>M</u>easure the size of its three angles. Use the calculator to display the sum of the three angles.

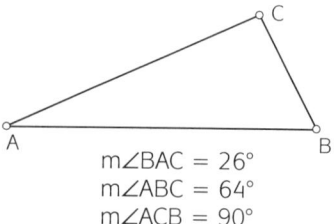

m∠BAC = 26°
m∠ABC = 64°
m∠ACB = 90°

1 What is the sum of the three angles of your triangle?

2 Select the vertex A. Using use the mouse, drag A to alter the size and shape of your triangle. Does the sum of the angles remain the same?

To measure angles, the arrow in the menu bar should be selected. Points need to be selected in the correct order.

3 Alter the size and shape of your triangle by dragging one of the other vertices. Is the sum of your three angles still the same?

Add the line BD to your triangle so that ABD is a straight line. <u>M</u>easure the size of the angle DBC and use the calculator to display the sum of the angles BAC and ACB.

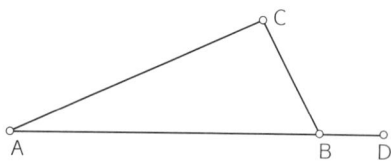

To ensure ABD is a straight line:
■ Select each line in turn
■ <u>M</u>easure their <u>S</u>lope
■ Adjust the slope of BD as necessary to make the slopes equal.

4 What do you notice?

Investigate what happens if you:

5 Drag the point B along the line AD.

6 Use the mouse to drag the vertex C to change your triangle.

Draw a quadrilateral ABCD and add the diagonal AC.

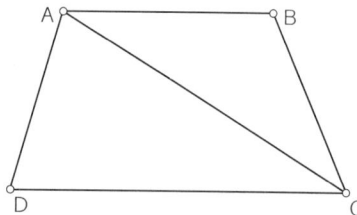

7 Explain why the sum of the four interior angles, DAB, ABC, BCD and CDA, of your quadrilateral is 360°.

8 Use the calculator to display the sum of your four interior angles and use the mouse to drag one of the vertices to confirm their sum is still 360°.

Note: When dragging a vertex, keep all angles in your quadrilateral less than 180°, since Geometer's Sketchpad does not measure reflex angles.

Activity sheet SSM8

Imagine

This activity sheet makes use of MSWLogo and the internet to take you to infinity by looking at area and perimeter.

The shapes below have a *finite* area and a *finite* perimeter

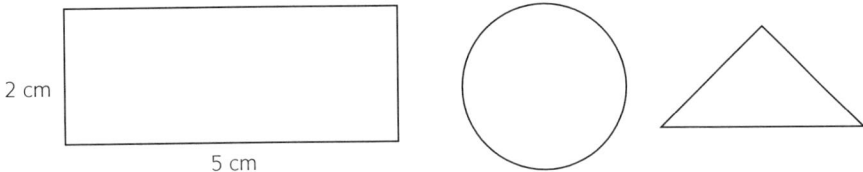

The *perimeter* of a shape is the distance all the way around its border.

The *area* of a shape is the amount of space the shape occupies.

The *rectangle* in Figure 1 has a perimeter of:
5 + 2 + 5 + 2 = 14 cm
and an area of:
5 × 2 = 10 cm².

Finite means that the areas and perimeters can be found ... they are not *infinite*!

Figure 1

Imagine a shape with a *finite area*, maybe no bigger than the rectangle above, but with an *infinite perimeter*! It is therefore impossible to walk all the way around its border. You could walk all year and make no progress!

To begin your journey to infinity, look at the 3 shapes below:

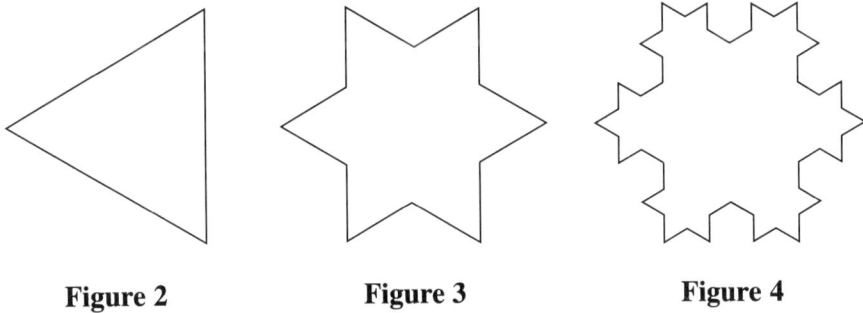

Figure 2   **Figure 3**   **Figure 4**

Figure 2 is an *equilateral triangle*. Each side of the triangle is then split into 3 equal sections and the middle section is transformed into another equilateral triangle with the base missing. This produces Figure 3. Each side of Figure 3 is split in a similar way to produce Figure 4. Imagine what happens if this continues for ever ... you will produce a shape with a *finite* area and an *infinite* perimeter.
Launch MSWLogo and load the file infinity.lgo ——————————— Click on <u>F</u>ile and then <u>L</u>oad

In the Commander box type:

    snow 0

Press the Enter key to produce the equilateral triangle in Figure 2.

Now enter snow 1 in the Commander box to produce the shape in Figure 3.

Enter snow 2 to produce the shape in Figure 4.

Continue your journey to infinity with:

snow 3, snow 4, snow 5, ...

The curves produced are called Koch snowflake curves, named after the Swedish mathematician Niels Fabian Helge von Koch.

(continued)

It is recommended that you go no higher than snow 8

You will have to let your imagination finish your journey!

You can use the internet to zoom in on some infinite perimeters. The *Mandlebrot* and *Julia* sets are a good source for this.

Look at:

http://www.wou.edu/las/natsci_math/math/burton/fractal.html

As you zoom in *anywhere* on the perimeter, see how the original Mandlebrot image is contained within the perimeter. Imagine!

Find out more on the Koch snowflake curve at:

http://www.agnesscott.edu/aca/depts_prog/info/math/riddle/ifs/ksnow/ksnow.htm

Find out more on Niels Fabian Helge von Koch at:

http://www-history.mcs.st-andrews.ac.uk/history/BiogIndex.html

The Mandlebrot and Julia sets are produced from *complex* functions. You can study complex numbers in 'A'-level Mathematics.

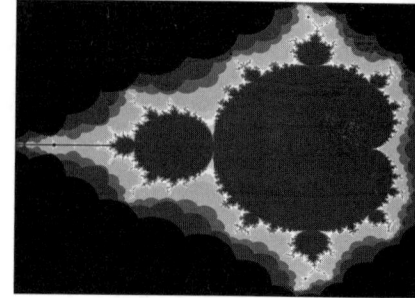

Part of the Mandlebrot set.

Coordinates, translations, reflections and rotations

This activity sheet looks at three different transformations. Before starting the activities launch Omnigraph.

Click on <u>S</u>hapes and then <u>F</u>lag to produce the flag ABCD with the base of the pole at the coordinate A(1, 1) (Figure 1).

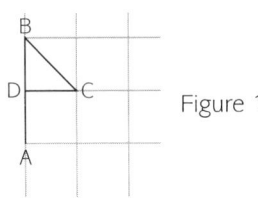
Figure 1

Translations

1 **(a)** Write down the coordinates of the other vertices B, C and D of the flag.

> B = (,) C = (,) D = (,)

(b) Write down the coordinates of the vertices A', B', C' and D' of the flag after a shift of 2 places to the right and 1 place down.

> A' = (,) B' = (,)
> C' = (,) D' = (,)

(c) Click on <u>T</u>ransform and then <u>T</u>ranslate and enter numbers in the *vector* on screen to check your answer.

(d) Use a vector to find A', B', C' and D' when the original flag is translated 3 places to the left and 2 places up.

> A' = (,) B' = (,)
> C' = (,) D' = (,)

(e) Write down the vector needed to slide the flag so that A' has the coordinates (–2, 6).

> The required vector is ( )

(f) Check your answer on screen.

(g) Translate the original flag so that A' has the coordinates
(i) (4, –1)   (ii) (1, 5)

> The required vectors are (i) ( ) (ii) ( )

Rectangle ABCD below has been moved 4 places to the *right* and 2 places *up*. A sliding movement like this is called a *translation*.

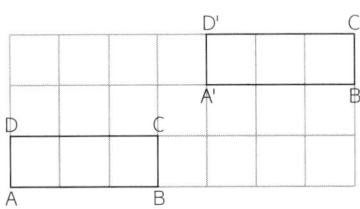

The vertices A,B,C, and D have moved to A',B',C', and D' respectively.

A mathematical way to describe the slide is by using

the *vector* $\begin{pmatrix} 4 \\ 2 \end{pmatrix}$ 4 places right
2 places up

The vector $\begin{pmatrix} -2 \\ -1 \end{pmatrix}$ would slide a shape 2 places *left* and 1 place *down*.

Use the 'hand' to help you shift the grid when necessary.

Reflections

Start a new window and produce the flag shown in Figure 1.

2 **(a)** Write down the new coordinates of the flag after a reflection in the y-axis.

> A' = (,) B' = (,)
> C' = (,) D' = (,)

(continued)

(b) Check your answer on screen by clicking <u>T</u>ransform and then <u>R</u>eflect and entering $x = 0$

$x = 0$ is alternative name for the y-axis.
Similarly, $y = 0$ is alternative name for the x-axis.

(c) Find A', B', C' and D' when the original flag is reflected in the x-axis.

> A' = (,) B' = (,)
> C' = (,) D' = (,)

(d) Find A', B', C' and D' when the original flag is reflected in the line (i) $x = 2$ (ii) $y = 1$ (iii) $y = x$

> (i) A' = (,) B' = (,)   (ii) A' = (,) B' = (,)
> C' = (,) D' = (,)   C' = (,) D' = (,)
>
> (iii) A' = (,) B' = (,)
> C' = (,) D' = (,)

Rotations

Start a new window and produce the flag shown in Figure 1.

3 **(a)** Write down the coordinates of the flag after a rotation about the point A, of:

(i) 90° clockwise   (ii) 90° anticlockwise
(iii) 180° clockwise   (iv) 180° anticlockwise

■ Click on <u>T</u>ransform then <u>R</u>otate and enter the details given.

Note:

■ For 90° *clockwise*, enter **–90**
■ For 90° *anticlockwise*, enter **90**
■ Change the angle for each new rotation – *you do not need to clear the screen.*

> (i) A' = (,) B' = (,)   (ii) A' = (,) B' = (,)
> C' = (,) D' = (,)   C' = (,) D' = (,)
>
> (iii) A' = (,) B' = (,)   (iv) A' = (,) B' = (,)
> C' = (,) D' = (,)   C' = (,) D' = (,)

(b) Comment on your answers to parts (iii) and (iv)

4 Start a new window.
■ Click <u>S</u>hapes, <u>P</u>entagon, and enter a length of 2 and centre (0,0) and press the Enter key.
■ Click on <u>T</u>ransform and then <u>R</u>otate.
■ Enter (0, 0) and then 60° – **do not** press the Enter key yet!
■ Click on <u>T</u>ransform and then <u>R</u>epeats and enter 50
■ Now press the Enter key to rotate the pentagon 50 times anticlockwise.
■ Rotate the pentagon 60° clockwise about the origin with 50 repeats.
■ Create some repeat rotations of your own with some different polygons.

Activity sheet SSM 10

Coordinates, enlargements and ratios

This activity sheet looks at enlargement, which is a type of transformation. Before starting the activities launch Omnigraph.

Click on <u>S</u>hapes and then <u>S</u>quare to produce a square with sides of length 2 and centre (1, 1).

1 **(a)** Write down the coordinates of the vertices A, B, C and D of the square:

> A = (,) B = (,)
> C = (,) D = (,)

(b) Write down the coordinates of the *image* A', B', C' and D' if ABCD is enlarged by a scale factor of 2 from the centre of enlargement C (0, 0):

> A' = (,) B' = (,)
> C' = (,) D' = (,)

(c) Click on <u>T</u>ransform and then <u>E</u>nlarge and enter the required numbers on screen to check your answer.

(d) Write down the coordinates of the image A' B' C' D' if ABCD is enlarged by a scale factor of 3 from the centre of enlargement (0, 0):

> A' = (,) B' = (,)
> C' = (,) D' = (,)

(e) Write down the coordinates of the image A' B' C' D' if ABCD is enlarged by a scale factor of 2 from the centre of enlargement (1, 1):

> A' = (,) B' = (,)
> C' = (,) D' = (,)

(f) Find the coordinates of the image A' B' C' D' if ABCD is enlarged by a scale factor of 2 from the centre of enlargement D (2, 0).

> A' = (,) B' = (,)
> C' = (,) D' = (,)

(continued)

The square ABCD below has been enlarged by a scale factor of 3 from the point A, to produce the square A' B' C' D'.

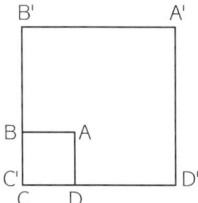

Each side of A' B' C' D' is *three* times longer than the corresponding side of the original square ABCD. The point C has not been moved, and C is in the same position as C'.

Use the 'hand' to help you shift the grid when necessary.

You can use Omnigraph to check your answers on screen.

(g) Write down the coordinates of the image A' B' C' D' if ABCD is enlarged by a scale factor of 2 from the centre of enlargement (1, 2).

```
A' = (  ,  ) B' = (  ,  )
C' = (  ,  ) D' = (  ,  )
```

(h) Write down the coordinates of the image A' B' C' D' if ABCD is enlarged by a scale factor of –2 from the centre of enlargement (1, 2).

```
A' = (  ,  ) B' = (  ,  )
C' = (  ,  ) D' = (  ,  )
```

(i) Write down the coordinates of the image A' B' C' D' if ABCD is enlarged by a scale factor of –3 from the centre of enlargement (0, 0).

```
A' = (  ,  ) B' = (  ,  )
C' = (  ,  ) D' = (  ,  )
```

2 (a) Use the square ABCD and its image A' B' C' D' to help you complete the following table.

Scale factor of enlargement	Ratio of lengths AB/A'B'	Ratio of areas ABCD/A'B'C'D'
2		
3		
4		
–2		
–3		
0.5		
–0.5		

(b) Comment on your findings.

Activity sheet SSM 11

Coordinates and straight lines

This activity sheet is about equations of straight lines. Before starting the activities launch Omnigraph.

1 Plot and join the two coordinates (2, 1) and (2, 7) and write down the coordinates of five other points on the line. Your line is part of the line called $x = 2$.

To draw a line in Omnigraph, select:
Shapes
and then
New shape

2 Plot and join the two coordinates (1, 4) and (5, 4) and write down the coordinates of five other points on the line [you do not need to use whole numbers]. This line is part of the line called $y = 4$.

Click this button to start a new question

3 Plot and join the two coordinates (–1, –5) and (–1, 3) and write down the coordinates of five other points on the line. Your line is part of the line called $x = –1$.

Use the hand button to move the grid as required.

4 Plot and join the two coordinates (4, –2) and (–2, –2) and write down the coordinates of five other points on the line. This line is part of the line called y = –2.

5 Start a new Window and select <u>A</u>nalysis and then <u>S</u>traight lines to draw the lines

   (a) $x = 3$

   (b) $y = 5$

   (c) $y = –4$

   (d) $x = –4$

   (e) Write down the coordinates at the points where any of the lines meet.

6 **(a)** Draw the line $y = x$ which passes through the origin and (4, 4).

   (b) Write down the coordinates of five other points on the line.

   (c) Draw the line $y = –x$ which passes through the origin and (–2, 2).

   (d) Write down the coordinates of five other points on the line.

7 Making use of the equal sign, give an alternative name for:

   (a) the x–axis

   (b) the y-axis

Patterns

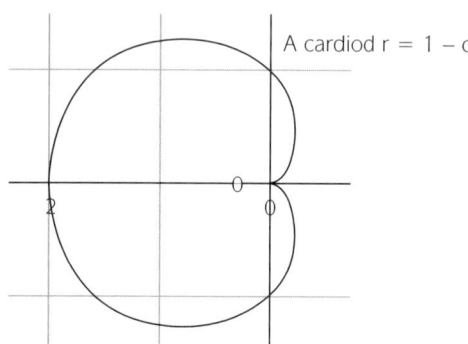

A cardiod $r = 1 - \cos\theta$

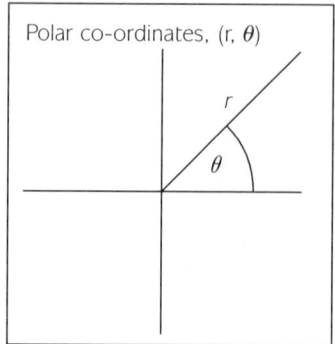

Polar co-ordinates, (r, θ)

This activity sheet uses a graphing package to help you use common trigonometric functions to create some unusual symmetrical shapes. It also introduces you to a new co-ordinate system. Before starting the activities, launch Omnigraph.

Investigate the family of curves:

1 $r = 1 - \cos n\theta$, for $n = 1, 2, 3, \ldots$

2 $r = 1 - \sin n\theta$, for $n = 1, 2, 3, \ldots$

3 $r = \cos n\theta + \sin m\theta$, for n and $m = 1, 2, 3, \ldots$

4 $r = \tan n\theta$, for $n = 1, 2, 3, \ldots$

5 $r = \sin^n\theta$, for n $= 1, 2, 3, \ldots$

6 $r = \cos^n\theta + \sin^m\theta$, for n and $m = 1, 2, 3, \ldots$

7 $r = \tan^n\theta + \cos^m\theta$, for n and $m = 1, 2, 3, \ldots$

8 $r = \frac{1}{1000}\theta$ for n revolutions, $n = 1, 2, 3, \ldots$

9 $r = \sin n\theta \tan m\theta$, for n and $m = 1, 2, 3, \ldots$

10 Investigate the behaviour of the curve $r = a + b\cos\theta$ if:
(a) + is changed to –
(b) $\cos\theta$ is changed to $\sin\theta$
Try different values of a and b such that $a < b$ and $a > b$

11 Investigate $r = \frac{1}{10}\theta^{\frac{1}{2}}$ by changing both the denominators and the number of revolutions.

12 Investigate $r = 4\theta^{-\frac{1}{2}}$ for 5 revolutions by changing the constant.

■ Discover many more mathematical curves like the 'Witch of Agnesi' at **http://www-groups.des.st-andrews.ac.uk/~history/Curves**

`r=1-cos`**2**`θ`   **for 1 revs**

You do not need to to clear the screen and type a new equation, just highlight the multiple of θ, change the number and press the Enter key.

$\sin^3\theta$ is the mathematical way to write $(\sin\theta)^3 \equiv \sin\theta \times \sin\theta \times \sin\theta$

Click:
this button to clear the curves as necessary:

this button to add powers:

this button to add fractions:

this button to add θ

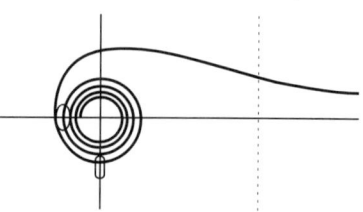

Activity sheet SSM13

Investigating straight lines 1

This activity sheet investigates the gradients of parallel and perpendicular lines. Before starting the activities you should launch The Geometer's Sketchpad.

Click on Display then Preferences and change the slope and calculation precision to hundredths.
Draw a straight line, AB, similar to the one shown.
Click Measure then Slope to measure the gradient of AB.
Use the point B to rotate the line.

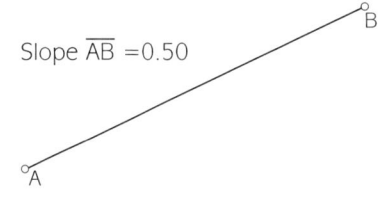

Slope $\overline{AB}$ =0.50

1 What is the difference between a line with a positive gradient and a line with a negative gradient?

2 What mathematical name is given to a line with zero gradient?

Rotate the line AB so that its gradient is 2.

3 What happens to the line as the gradient increases from 2 and stays positive?

4 What would be the gradient of a vertical line?

Rotate the line AB so that its gradient is 2 again.
Place a point C to the right of the line. Select the point C together with the line AB and Construct a Perpendicular Line, k. Measure the gradient of the line k.

Use the Calculator to display the product of the two gradients and rotate the two perpendicular lines by dragging either of the points A or B.

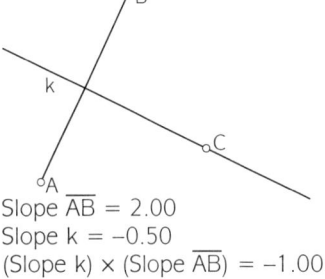

Slope $\overline{AB}$ = 2.00
Slope k = −0.50
(Slope k) × (Slope $\overline{AB}$) = −1.00

5 What is the connection between the gradients of perpendicular lines?

Draw a new line DE and add a point F to the right of the line. Select the line and the point F and Construct a Parallel Line with slope m.

6 What do you notice about the gradients of parallel lines?

The line $y = mx + c$ passes through the y-axis at the point $(0, c)$ and has gradient m.

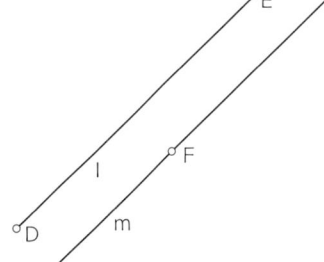

Open a new sketch and click Graph, Show Axes.

7 On the same set of axes draw the lines
  (a) $y = 2x + 1$
  (b) $y = 3x - 2$
  (c) $y = -4x + 2$
  (d) $y = \frac{1}{2}x + 1$
  (e) $y = -\frac{1}{2}x + 1$

Hint:
Rotate the line to the correct gradient then slide it into the correct position.

Investigating straight lines 2

This activity sheet shows you how to produce straight lines by a series of transformations. Before starting the activities, launch Omnigraph.

1 Enter the line with equation $y = x$.
The line passes through $(0, a)$ and $(1, b)$. What are the values of a and b?

2 Enter the line with equation $y = 2x$.
The line passes through $(0, c)$ and $(1, d)$. What are the values of c and d?

The line $y = x$ has been transformed to produce the lines $y = 2x$ and $y = 3x$.

3 Enter the line with equation $y = 3x$.
The line passes through $(0, e)$ and $(1, f)$. What are the values of e and f?

4 What is the equation of the line which passes through
   (a) $(0, 0)$ and $(1, 4)$?
   (b) $(0, 0)$ and $(1, 5)$?
   (c) Enter the equations of your two lines to check your answers.

5 Enter the line with equation $y = \frac{1}{2}x$.
The line passes through $(0, g)$ and $(h, 1)$. What are the values of g and h?

6 Enter the line with equation $y = \frac{1}{4}x$.
The line passes through $(0, k)$ and $(l, 1)$. What are the values of k and l?

7 Copy and complete this table.

Equation of the line	Passing through
$y = 8x$	$(0,\underline{\ \ })$ and $(1,\underline{\ \ })$
$y = \underline{\ \ }$	$(0,0)$ and $(1,6)$
$y = \frac{1}{3}x$	$(0,\underline{\ \ })$ and $(\underline{\ \ },1)$
$y = 10x$	$(\underline{\ \ },0)$ and $(\underline{\ \ },10)$
$y = \underline{\ \ }$	$(0,0)$ and $(5,1)$

Clear the screen before each of the following questions.

8 Enter the lines with equations:
   (a) $y = x$
   (b) $y = x + 1$
   (c) $y = x + 2$
   (d) $y = x - 3$
   (e) $y = x - 5$
   (f) Write down the coordinate where each line crosses the y-axis.

Each line in Question 8 **(b)** to **(f)** is parallel to the line $y = x$ and is a *translation* [shift] from $y = x$ parallel to the y-axis.

(continued)

9 Write down the equations of five more lines parallel to the line $y = x$ and check them on screen.

10 Write down the equations of five lines parallel to the line $y = 2x$ and check them on screen.

11 Write down the equations of five lines parallel to the line $y = 4x$ and check them on screen.

12 Enter the lines with equations $y = 2x + 1$ and $y = -2x + 1$. The lines are *reflections* of each other in the *y*-axis.

13 For each of the following equations, enter the line and its reflection in the *y*-axis.

Clear the screen before each part of the question.

  (a) $y = 5x + 1$

  (b) $y = 3x - 4$

  (c) $y = -4x + 2$

  (d) $y = \frac{1}{3}x - 1$

  (e) $y = -\frac{1}{2}x + 1$

14 On graph paper, draw the *x* and *y* axes from –8 to + 8, using 1 cm to represent 1 unit on each axis.

  (a) Draw the line $y = x$ on your graph paper.

Use each line that you draw to help you work out how to draw the next line.

  (b) Add the line $y = 3x$

  (c) Add the line $y = 3x - 2$

  (d) Add the line $y = -3x - 2$

15 On a new pair of axes identical to those in Question 14, draw the line $y = -2x + 4$ in four stages, starting with the line $y = x$.

16 On a new pair of axes identical to those in Question 14, draw the line $y = -\frac{1}{4}x - 2$ in four stages, starting with the line $y = x$.

Activity sheet SSM 15

Drawing bearings

This activity sheet asks you to draw points of the compass and bearings. Before starting the activities, launch MSWLogo.

Bearings are measured from the north in a clockwise direction.

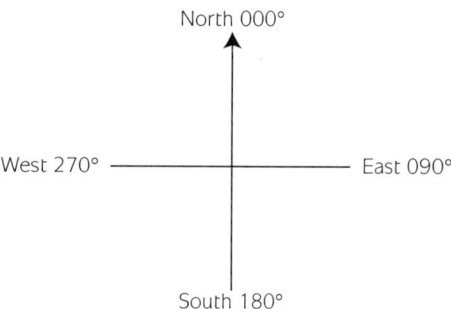

In Questions 1 and 2 make the arms of your drawings a length of 100 units.

1 Draw a bearing of:
   (a) 040°
   (b) 130°
   (c) 210°
   (d) 335°

It is a good idea to sketch your bearings on paper first to help you work out the angles to turn through.
To draw a bearing of 150°, type
   fd 100
   label "North
   bk 100
   rt 150
   fd 100

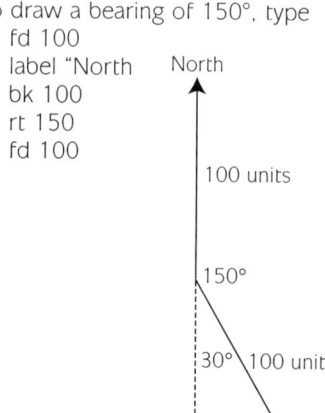

2 Draw a diagram with labels to represent the compass rose:

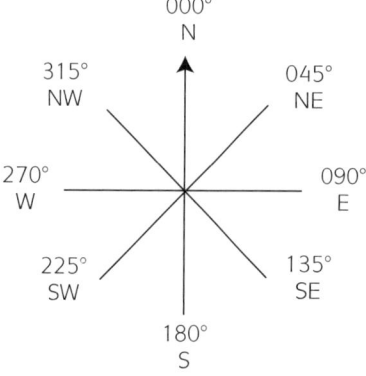

South East is halfway between South and East, a bearing of 135° (90° + 45°)

3 In this question, let 1 unit in MSWLogo represent 2 km. Draw a diagram to represent a journey of:
   (a) 270 km East followed by 220 km South West
   (b) 180 km on a bearing of 130° followed by 310 km on a bearing of 075°
   (c) 215 km West followed by a journey of 200 km at a bearing of 300°

4 Try creating some journeys of your own.

Investigating circles 1

This activity sheet investigates tangents and chords of a circle. Before starting the activities launch The Geometer's Sketchpad.

■ Draw a circle and select the centre and the point on the circumference.
■ Construct a Segment to obtain a radius.
■ Select the radius and the point on the circumference and Construct a Perpendicular Line to create a *tangent* to the circle at your point on the circumference.
■ Add a point C to the line.

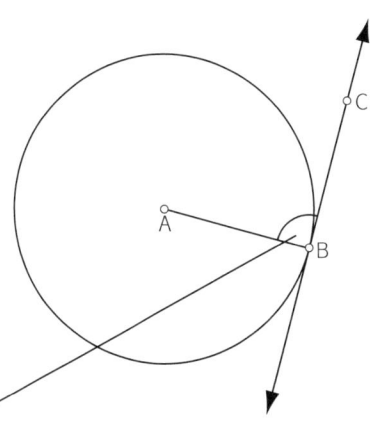

m∠ABC = 90°

1 Measure Angle ABC.

Use the mouse to drag the point A so that the circle changes size and the gradient of the tangent changes.

2 What can you say about the angle between a tangent at a point on the circumference of a circle and the radius at that point?

On the same sketch draw a second circle.

■ Add a point F, on the circumference.
■ Select both points, E and F and Construct a Segment to obtain the *chord*, EF.
■ Select the chord and the centre of your circle and Construct a Perpendicular Line.
■ Add a point G where this perpendicular line intercepts your chord, as shown.

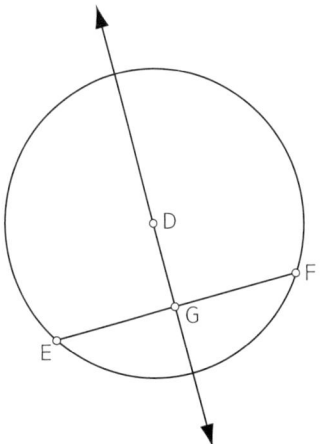

3 Measure the Distances EG and GF.

Drag the point D at the centre of your circle and monitor the distances EG and GF.
What can you say about the perpendicular line through the centre of your circle and your chord?

Activity sheet SSM17

Investigating circles 2

This activity sheet investigates the properties of circles.
Before starting the activities launch The Geometer's Sketchpad and
use Display Preferences to change the angle precision to tenths.

1 Draw a circle and construct the quadrilateral ABCD.
Measure Angles CAB and CDB.
Investigate the connection between the two angles as you drag
the point B round the circumference.

1

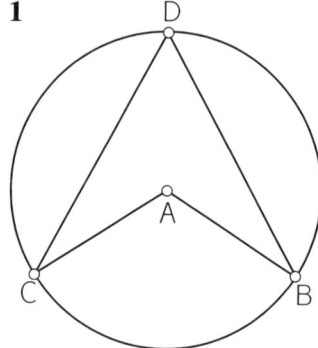

2 ■ Draw a line EF and Construct a Point at Midpoint.
■ Select point G then E and Construct a Circle By Centre and
Point.
■ Add point H to the circumference and construct segments EH
and HF.

Measure Angle EHF.
Investigate the size of angle EHF as you drag the point H round
the circumference.

2

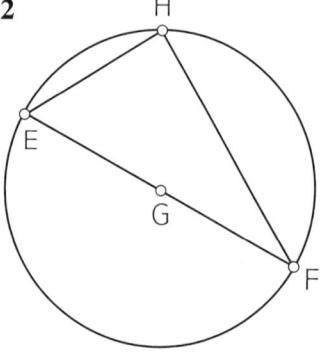

3 Draw a third circle and construct the chord LK and the triangles
LKN and LKM.
Measure Angles LNK and LMK.
Investigate the size of these two angles as you drag the point K
round the circumference.

3

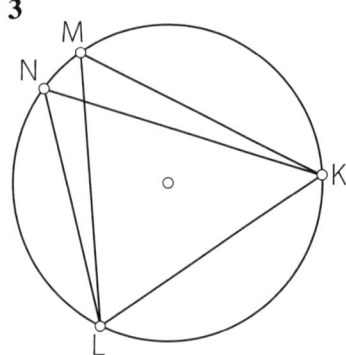

(continued)

4 Draw a fourth circle and construct the *cyclic quadrilateral* PQRS.
Measure Angles PSR and PQR.
Use the calculator to display their sum.
Investigate their sum as you drag any of the points P, Q, R and S round the circumference.

4

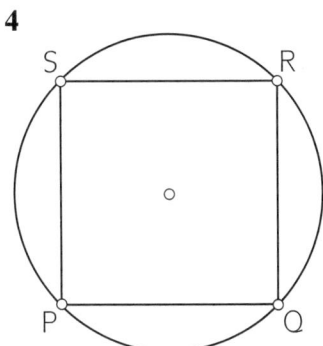

5 Draw a final circle and construct a radius UV.
To create a tangent, Construct a Perpendicular Line to UV.
Construct the triangle VWX and add a point Y on the tangent.
Measure Angles YVW and VXW.
Investigate their size as you drag the point W round the circumference.

5

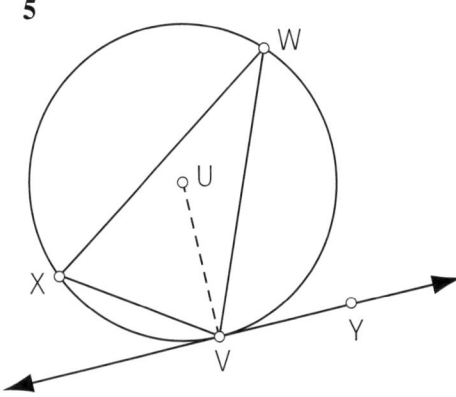

6 Write down all your findings.

Triangles, bisectors and circles

This activity sheet investigates the different circles obtained when you bisect the angles and sides of a triangle. Before starting the activities you should launch Geometer's Sketchpad.

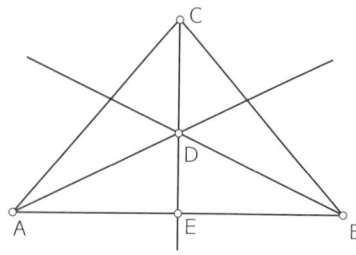

- Draw a line AB.
- Add a point C above AB.
- Select the points A and C and Construct a line Segment AC.
- Select the points B and C and Construct a line Segment BC.
- Select angle CAB and Construct an Angle Bisector.
- Bisect the angle ACB in the same way.
- Add a point D where the two bisectors intercept.
- Bisect angle ABC.
- Add a point E where the bisector of angle ACB crosses the line AB.
- Select the point D then the point E and Construct a Circle By Centre And Point.

1 Describe the circle constructed.

2 Open a new sketch.

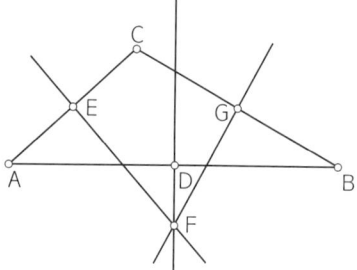

- Construct a second triangle ABC.
- Select the line AB and Construct a Point At Midpoint.
- Select the line AB and the point at its centre and Construct a Perpendicular Line.
- Construct a line perpendicular to AC.
- Add the point, F, where the two perpendicular lines intercept.
- Construct a line perpendicular to BC.
- Select F and the vertex C of the triangle and Construct a Circle By Centre And Point.

Describe the circle constructed.

3 Click Edit and Undo Construct Circle.
Select F again and then the vertex B of the triangle and Construct a Circle By Centre And Point.

Describe the circle constructed.

4 Repeat the above instructions but this time use F and the vertex A of the triangle.

Describe the circle constructed.

Investigating geometrical sequences

This activity sheet uses a spreadsheet to help you investigate the limit of a geometric sequence of circles and polygons. The activity makes use of a calculator, trigonometry, angles in polygons in both degrees and radians, limits and formula creation.

You can see a mathematical proof of this investigation by Tamara Curnow in the journal: *Mathematical Spectrum*, Volume 26, Number 4

The circle opposite has a radius of 10 cm. Inscribed in the circle is an *equilateral* triangle, ABC. The angle formed by points A, C and the centre of the circle, O, is 120°, so AOD = DOC = 60°.

1 Use a calculator and trigonometry to calculate the length of OD.

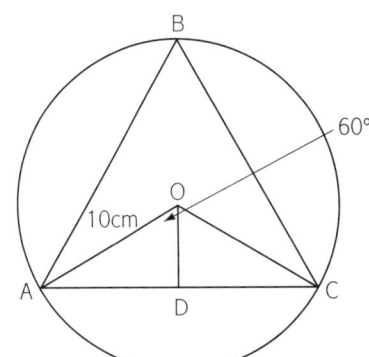

2 A circle of radius OD is inscribed in the triangle and a square is inscribed in the new circle.
(a) Write down the length of the radius OE.
(b) Calculate the size of the angle EOF
(c) Use trigonometry to calculate the length of OF.

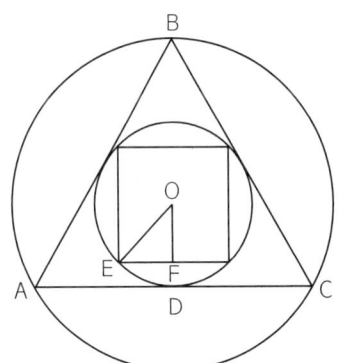

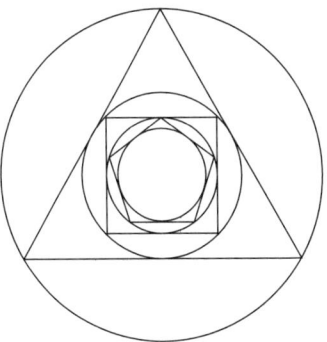

The sequence continues:
■ A circle is inscribed inside the square.
■ A regular pentagon is inscribed inside the new circle.
■ A circle is inscribed inside the pentagon.
■ A regular hexagon is inscribed inside the new circle.
■ A circle is inscribed inside the hexagon and so on...

The radius of each new circle is the previous radius multiplied by the cosine of the angle, *x*, at the top of each new right-angled triangle.

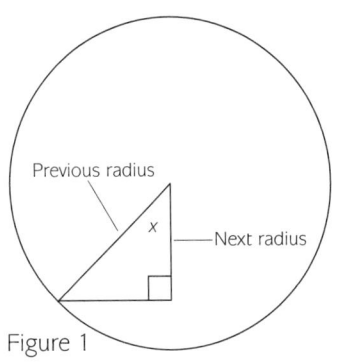

Previous radius

x
Next radius

(continued)   Figure 1

Launch Excel and set up a spreadsheet similar to the one below:

	A	B	C	D	E	F
1	Number of sides of the polygon	Angle x, in degrees. See Figure 1	Angle x, in radians	Next radius	Starting radius	Ratio of 'final' radius to starting radius
2	3	=180/A2	=B2*pi()/180	=E2*cos(C2)	10	
3	=A2+1	=180/A3	=B3*pi()/180	=D2*cos(C3)		
4	=A3+1	=180/A4	=B4*pi()/180	=D3*cos(C4)		

	A	B	C	D	E	F
1	Number of sides of the polygon	Angle x, in degrees. See Figure 1	Angle x, in radians	Next radius	Starting radius	Ratio of 'final' radius to starting radius
2	3	60	1.047197551	5	10	
3	4	45	0.785398163	3.535533906		
4	5	36	0.628318531	2.860307014		
5	6	30	0.523598776	2.477098537		
6	7	25.71428571	0.448798951	2.231788664		
7	8	22.5	0.392699082	2.061903868		

- Highlight cells A7 to D7 and drag this square down to row 500.
- In cell F2 enter the formula =D500/E2, to calculate the ratio of the 'final' radius to the starting radius.
- Choose some different values for the starting radius and comment on the values produced in cells F2 and D2.
  To match the value of 0.114959357 in Tamara Curnow's proof you need a polygon of about 33000 sides!

Notes on Excel:
When you want to use a trigonometric ratio like cosine, the angle must be in *radians* and not *degrees*.
π can be entered as pi()
Use =cos(C2) to find the cosine of the angle in cell C2.

Remember:
To turn an angle from degrees into radians:
- multiply by π and then divide by 180

Extension

Set up a spreadsheet to investigate what happens if:

- The polygons are built *outside* the circles.

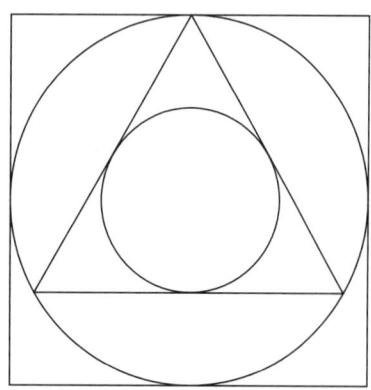

- The sequence is sphere, cube, sphere, cube, ... This makes use of Pythagoras' theorem in three dimensions.

Approximating a value for π

This activity sheet uses Pythagoras' theorem and area to estimate a value for π.

Remember that π is an *irrational* number and it is therefore not possible to find an exact value for it.

Using a calculator

1 The circle shown has a radius of 5 units. Use Pythagoras' theorem to calculate the lengths y_1, y_2, y_3 and y_4 accurate to 3 decimal places.

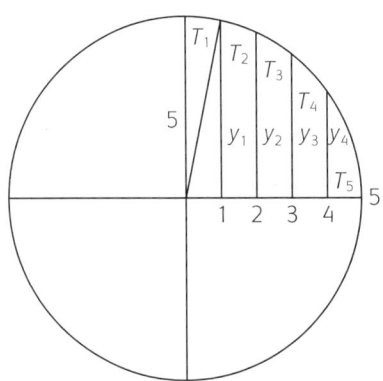

2 Calculate the area of the approximate *trapeziums*, T_1, T_2, T_3, T_4 and the approximate *triangle* T_5.

3 Calculate the approximate area of the *quadrant*.

4 The area of this quadrant is $\frac{1}{4}$ the total area of the circle, $\pi \times 5 \times 5 = 25\pi$.
Equate your approximate area of the quadrant to 25π and rearrange your equation to obtain an approximate value for π.

Using a spreadsheet

Before starting these activities, launch Excel.

5 Use a spreadsheet to calculate the approximate area of the quadrant, this time split into 100 parts, containing 99 approximate trapeziums and 1 approximate triangle. Use this area to obtain a more accurate approximate value for π, accurate to 9 decimal places.

It is a good idea to plan your spreadsheet on paper first. Use:
- Column A for x values – the perpendicular distances between the trapeziums
- Column B for y values
- Column C for areas

6 An ellipse has an area πab, where a and b are the lengths shown. If $a = b$ the ellipse will be a circle. Splitting the ellipse into 1000 parts, find an approximate value for π, if:
(a) $a = 10$ and $b = 5$ (b) $a = 5$ and $b = 10$

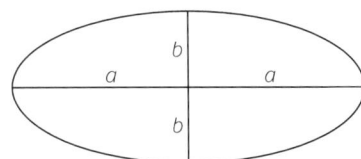

7 Comment on the accuracy of your three spreadsheet approximations compared with a calculator value for π of 3.141592654

Investigating trigonometric curves

This activity sheet investigates the connections between trigonometric curves, making use of transformations. Before starting the activities, launch Omnigraph.

1 On the same set of axes enter the curves:
$y = \sin x$ and $y = \cos x$.
   (a) Write down any similarities and connections between the two curves.
   (b) At what values of x does each curve cross the x-axis?
   (c) At what values of x does each curve reach its maximum and minimum values of ± 1?

Before entering any equations:
Select
■ Zoom and Trig scales
followed by
■ Zoom and Rescale the x-axis
from −720° to 720°

2 On the same set of axes add the curve $y = \tan x$.
   (a) What are the maximum and minimum values of $\tan x$?
   (b) At what values of x do these maximum and minimum values occur?
   (c) At what values of x does the curve $y = \tan x$ cross the x-axis?

3 Use your three curves to help you complete this table:

curve \ angle	0° < x < 90°	90° < x < 180°	180° < x < 270°	270° < x < 360°
$y = \sin x$	positive			
$y = \cos x$			negative	
$y = \tan x$				negative

4 What is the single transformation from $y = \sin x$ to $y = -\sin x$?

Clear the screen before each part of the following question.

5 From the list:
$y = \pm \sin x, y = \pm \cos x$ and $y = \pm \tan x$
state what each of the following curves are equivalent to:
   (a) $y = \cos(90° - x)$ (b) $y = \sin(90° - x)$
   (c) $y = \sin(180° - x)$ (d) $y = \cos(180° - x)$
   (e) $y = \cos(360° + x)$ (f) $y = \sin(360° + x)$
   (g) $y = \tan(180° - x)$ (h) $y = \tan(360° + x)$
   (i) $y = \sin(-x)$ (j) $y = \cos(-x)$
   (k) $y = \tan(-x)$

6 Investigate the transformations between:
   (a) $y = \sin x$ and $y = \sin ax$, if $a > 1$
   (b) $y = \sin x$ and $y = \sin ax$, if $0 < a < 1$
   (c) $y = \cos x$ and $y = b\cos x$, if $b > 1$
   (d) $y = \cos x$ and $y = b\cos x$, if $0 < b < 1$
   (e) $y = \sin x$ and $y = \frac{1}{2}\sin 3x$
   (f) $y = \cos x$ and $y = 2\cos \frac{1}{3}x$

Activity sheet HD1

Mathematics through the ages

This activity sheet will help you find information about famous mathematicians by using the internet.

Go to the internet site:

http://www-history.mcs.st-andrews.ac.uk/history/BiogIndex.html

and use it to help solve the following crossword.

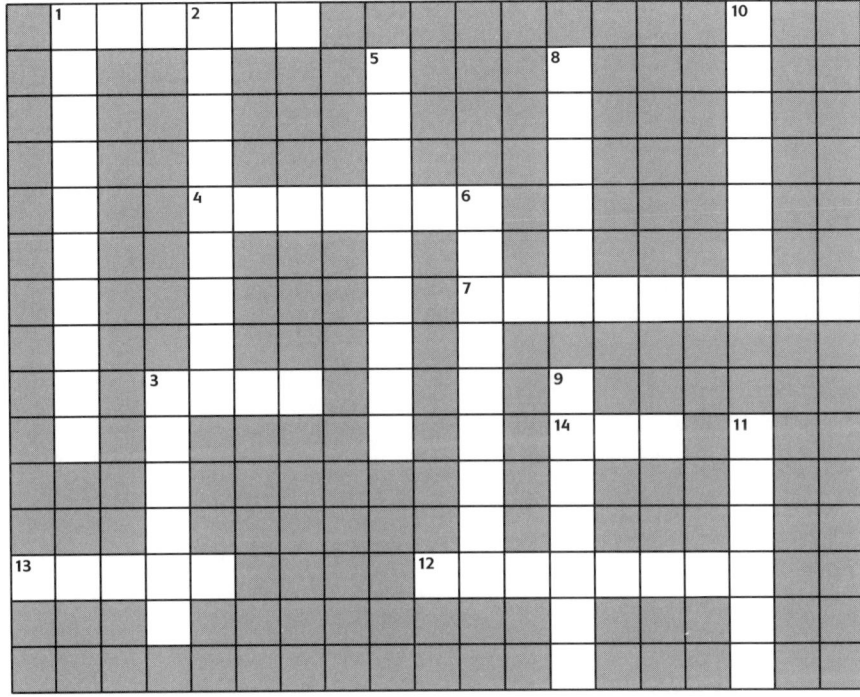

For each mathematician in the crossword make a note of:
- their date of birth and death
- two other mathematicians linked with them

ACROSS

1 Had a triangle of numbers named after him. His dad was a mathematician and lawyer.
3 Heron was often called this. His formula can be spotted on pencil cases.
4 His surname is similar to the name of one of Santa's reindeer.
7 Surname of a family of 2^3 Swiss mathematicians.
12 I came up with the formula: $E = mc^2$
13 Forename of 'Ouch, that apple hurt!'
14 I'm electric at mathematics.

DOWN

1 Famous for his theorem on right-angled triangles.
2 The University where $19 - 3 \times 2$ Across taught.
3 Mother of the Englishman who discovered calculus the same time as Leibniz.
5 Scottish mathematician whose series, like Taylor's series, is studied by A-level students.
6 This Italian had a sequence of numbers based on rabbits.
8 Best known for his impossible drawings.
9 Real name of 'the fourth triangle number' Down.
10 Wrote stories about Alice.
11 Nationality of $4^2 + 1 - \sqrt{256}$ Across.

Now try the following activities:
- List the Mathematicians in the crossword in chronological order.
- Calculate their mean average age at death.
- Write down the first 7 rows of the triangle named after 1 Across.
- Find the 'sequence of rabbits' in your triangle.

A day out in London

This activity sheet asks you to find information from the internet and make use of the 24-hour clock and the calendar.

Use the internet to plan a day out in London for 2 adults and 3 children under 12. You have a budget of £300.
Plan your day carefully and make a note of all times, prices and locations. You could use a spreadsheet to help keep a running total.

Have a nice day!

If you are living outside of London, travel to London by National Express Coach.

For travel details log on to
  http://www.gobycoach.com
Use the Around Britain with National Express link, then Times and Fares link for bus times and prices.

When in London you will need a One-day Travel Card each.
Log on to:
  http://www.londontransport.co.uk/ftt_oneday.shtml
for the card prices.

Go and see a film at the Prince Charles Cinema.
Choose one at:
  http://www.princecharlescinema.com/
and find out the cost of tickets with the *Films from £1.50 only at the PCC* link.

Take in a show:
  http://www.londontheatre.co.uk/

Finally, go for a meal:
  http://www.geocities.com/restaurantamici/
  http://freespace.virgin.net/cathy.white/
  http://www.personal.u-net.com/~jubjang/sweet.htm
  http://www.1lombardstreet.com/
  http://www.putneybridgerestaurant.com/

Flights

This activity sheet asks you to find information from the internet and makes use of the 12- and 24-hour clock and the calendar.

Log on to the British Airways internet site at:

http://www.british-airways.com/

Enter *today's* day, date and time:

Day	Date	Time

1 For each of the flights in the table below, use the Schedules link to find:
- the departure time
- the arrival time, including any day changes
- the length of journey, including any times between changes of planes

After entering the cities of departure and arrival, use the *List Flights* link to obtain the timetables.

City of departure	City of arrival	Instructions	Departure time	Arrival time	Length of journey	Connecting flight
Gatwick	Frankfurt	Next flight				
Manchester	Madrid	Earliest flight tomorrow				
Gatwick	Warsaw	Last flight today				
Glasgow	Gatwick	Next flight				
Manchester	Toronto	Earliest flight next Saturday				
Heathrow	Sydney	Next flight				
Singapore	Adelaide	Earliest flight tomorrow				
Heathrow	Dubai	Next flight				
Glasgow	Lisbon	Earliest flight tomorrow				
Moscow	Heathrow	Last flight today				

2 Change all the 24-hour times to the 12-hour clock.

3 Give the day and date of arrival in each city.

Activity sheet HD4

Looking at data 1

This activity sheet involves using a spreadsheet and includes creating formulae, mean averages, sorting, analysing data, scatter graphs and correlation. Before starting the activities you need to launch Excel and open Soccer League Tables.xls .

1 **(a)** In cell L2 enter a formula to calculate the total number of points gained by Birmingham City. Copy down your formula to cell L25 to obtain the total of points obtained by each team.

Use this data in cells B2, C2, G2 and H2 with 3 points for a win and 1 point for a draw.

(b) Enter a formula in cell M2 to generate a formula to calculate the total number of goals scored for each team. Enter a formula in cell N2 to calculate the total number of goals conceded by each team. Copy the formulae down to calculate the totals for each team.

(c) In cell O2 enter a formula to calculate the goal difference for Birmingham City. Copy down this formula to obtain the goal difference for each team.

Goal difference is the total number of goals scored minus the total number of goals conceded. Use the data in cells E2, F2, J2 and K2.

(d) Highlight the cells from A2 across to O2 down to row 25 and sort the cells in descending order by the points scored.

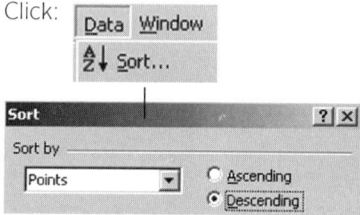

Click:

Use the formula:= Average(B2:B25)

(e) Make a note of the bottom three teams who were relegated that season.

2 **(a)** In cell B26 enter a formula to calculate the mean number of Home wins for the season. Copy the formula across to cell K26 to calculate the other mean averages.

(b) Comment on why some of these means are the same.

Use the Σ button on the main toolbar to check your answer.

3 What do you think the total of the numbers in the Goal Difference column should be?

Use the Ctrl key on your keyboard to highlight columns that are not next to each other.

4 Highlight cells LM to L25 and cells M1 to M25. Create a scatter graph of your data and comment on it.

5 Use columns M and N to create a scatter graph showing the points total against the number of goals conceded by each team. Comment on the graph.

6 Is there any correlation between the total number of goals scored and the total number of goals conceded for each team?

7 **(a)** Before the 1981/82 season only 2 points were awarded for a win. Change your formula in cell L2 and copy it down to see what the points total would have been prior to 1981/82.

(b) Sort the teams in descending order using this new points total.

(c) Highlight the three rows containing the data of the teams with 36 points. Sort these three rows in descending order using the Goal Difference column.

(d) Make a note of the bottom three teams who would have been relegated if only 2 points were awarded for a win. Which two teams were affected most by the points change?

Looking at data 2

This activity sheet asks you to use a spreadsheet to describe social trends, interpret data, work out percentages and make predictions. Before starting the activities launch Excel.

Open the spreadsheet file olympics.xls.

1 **(a)** In the Olympic Games of 1904 and 1932 there were far fewer competitors than in the games just before and just after. Why do you think that was?

(b) The Olympic games have been held every 4 years since 1896 apart from which years?

(c) Why were there no Olympic games in those years?

(d) The USA and USSR have been the most successful countries in gaining the most gold medals. Why do you think that is?

(e) Why was the USSR not successful before 1956?

(f) In 1980 and 1984 the USSR and the USA, respectively, gained many more gold medals than at any other time. Why do you think that was?

2 **(a)** Highlight the data in cells E2 down to cell E24. Drag the black square at the bottom right of cell E24 down to about row 40. In which year does this predict the Olympic time for the Men's 400 m will drop below 40 seconds?

Highlight cells A22 to A24 and drag down to row 40 to display further years

(b) Use a similar method to predict the year when the Women's 400 m time will drop below 40 seconds.

3 Enter a formula in cell J2 and drag it down to cell J24 to calculate the percentage of female competitors.

(a) In which years did the percentage fall?

(b) Use the Ctrl key to select cells A2 toA24 and J2 to J24. Produce a scattergraph for this data and finish the chart wizard by selecting ...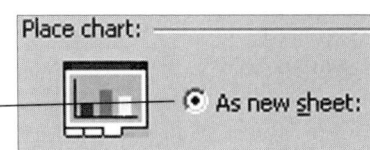

(c) Change the scale of the *y*-axis to a maximum of 60 and the scale on the *x*-axis to a maximum of 2040, print out your graph and use it to predict the year in which the number of female competitors will reach 50%.

4 Use column G to predict the year when a jump of 9 feet will be needed to win the Men's high jump.

Activity sheet HD6

Business trip

This activity sheet uses the internet to help you plan a business trip. It makes use of the 24-hour clock and currency conversion.

Use the following four internet sites to find the departure and arrival times for the trains and planes needed to plan the business trip shown in the table below. You need to catch the *next* train from Folkestone Central.

Enter *today's* day, date and time:

Day	Date	Time

For trains in England – use the **timetable** link at:
> **http://www.railtrack.co.uk/**

For flights from England to the USA and from Canada back to England – use the **Schedules** link at:
> **http://www.british-airways.com/**

For trains in the USA – use the **Schedules and sample fares** link at:
> **http://reservations.amtrak.com/**

For flights in the USA – use the **View Schedules** link at:
> **http://www.continental.com/**

From	To	Departure time	Arrival time	Length of journey	Day	Date	Transport	Details
Folkestone Central	Gatwick Airport						Train	Catch next flight
Gatwick Airport	New York						Plane	Stay 2 nights
New York	Boston						Train	Stay overnight
Boston	Los Angeles, CA						Plane	Stay 2 nights
Los Angeles, CA	Montreal, PQ						Plane	Stay 4 nights
Montreal, PQ	Heathrow						Plane	Stay overnight
Heathrow	Folkestone Central						Train	End of trip

- Find the total amount of time you will spend travelling.
- You allow £5600 for your trip. Use the currency converter at:
  http://finance.yahoo.com/m3?u
  to change £4800 into *US* dollars and the *rest* into *Canadian* dollars.
- At the end of the trip you have 255 US dollars and 320 Canadian dollars left. Convert these back into British pounds.
- Use the distance calculator at:
  http:// www.record–eagle.com/lib/distan.htm to work out the total distance travelled in North America.

Activity sheet HD7

Cumulative frequency

This activity sheet makes use of a spreadsheet and helps you to estimate medians, upper and lower quartiles and interquartile ranges from cumulative frequency curves. Before starting the activities, launch Excel.

1 This table shows the results of a speed check on 200 vehicles. The speeds were rounded to the nearest 10 kmh⁻¹.

10 kmh⁻¹ means 10 kilometres per hour.

Speed	40	50	60	70	80	90	100	110	120
Number of vehicles	2	3	14	16	33	47	52	22	11

(a) Enter the data into columns A and B of a spreadsheet and use column C to calculate the cumulative frequencies of the data.

	A	B	C
1	**Speed**	**Frequency**	**Cumulative frequency**
2	40	2	2
3	=A2+10	3	=C2+B3

(b) Using the Ctrl key highlight the cells in columns A and C from row 1 down to row 10. Use the chart wizard to create a cumulative frequency curve using the

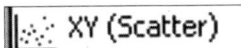 XY (Scatter)

option at Step 1, adding labels at Step 3 and finish by selecting a new sheet at Step 4.

(c) Double click on the *vertical* axis of your graph and change the major unit to 10 and in a similar way change the major unit on the *horizontal* axis to 5. Print out a copy of your cumulative frequency curve and use it to find estimates for:
  (i)   the median speed
  (ii) the interquartile range
  (iii) the number of vehicles travelling at more than 85 kmh⁻¹
  (iv) the number of vehicles travelling at less than 65 kmh⁻¹

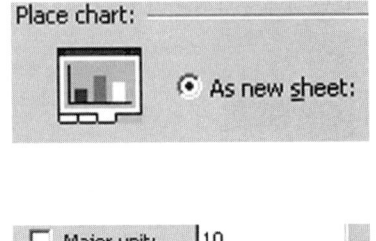

(continued)

2 The table below shows the intervals when goals were scored in normal time during all the games in Euro2000.

■ Euro2000 was the European Nations soccer tournament held in Belgium and Holland during June 2000.
■ Normal time does not include extra time.

Time interval in minutes	Tally	Frequency
1–15	THL |||	
16–30	THL ||||	
31–45	THL THL THL ||	
46–60	THL THL THL THL |||	
61–75	THL THL |||	
76–90	THL THL ||	

(a) Use the Tally column to complete the Frequency column.

(b) Which time interval contains the modal number of goals?

(c) Use formulae to enter the time intervals into columns A and B of a spreadsheet and use the table to complete column C

	A	B	C	D
1	Time in minutes from	Time in minutes to	Frequency	Cumulative frequency
2	1	15	8	8
3	=A2+15	=B2+15	9	=D2+C3

(d) Use a formula in column D to generate the cumulative frequencies.

(e) What was the total number of goals scored in normal time?

(f) Use the Ctrl key to highlight the data in columns B and D and use the chart wizard to create a cumulative frequency curve.

Remember:
■ Use the points at the upper end of the range when plotting points from a *grouped* frequency table.
■ Add labels to the axes and a title to your curve.
■ Place the chart as a new sheet.

Change the major unit on the vertical axis to 3 and on the horizontal axis to 5 and print out a copy of your cumulative frequency curve.

(g) Use your curve to estimate:
    (i) the number of goals scored *after* 55 minutes
    (ii) the number of goals scored *before* the 75th minute

Activity sheet HD8

Quartiles, averages and ranges

This activity sheet uses a spreadsheet to help you calculate lower and upper quartiles, interquartile ranges, means and medians. It includes skew and box plots. Before starting the activities, launch Excel.

1 Open the file houseprice.xls. The data, from May 2000, shows the price of an average semi-detached house in various parts of England, Wales and Northern Ireland.

(a) Highlight the data in the two columns from row 2 down to row 21, click <u>D</u>ata then <u>S</u>ort... and sort the data in *ascending* order by price.

(b) Enter a formula in cell B22 to calculate the *range* for the house prices.

> The range of a set of data is the highest value minus the lowest value.

(c) In cell B23 enter the formula =average(B2:B21) to calculate the *mean* house price.

(d) In cell B24 enter the formula =quartile(B2:B21,1) to calculate the *lower quartile* house price.

(e) In cell B25 enter the formula =quartile(B2:B21,2) to calculate the *median* house price.

(f) In cell B26 enter the formula =quartile(B2:B21,3) to calculate the *upper quartile* house price.

(g) In cell B27 enter a formula to calculate the *interquartile range* for the house prices.

(h) Explain why there is such a large difference between the *mean* and *median* averages.

(i) Discuss which of the two types of average you think the Government would choose to use.

(j) On paper, draw a *box plot* and explain, with reasons, whether the data has a positive or a negative *skew*.

> Some graphical calculators allow you to draw box plots.
> Another name for a *box plot* is a *box-and-whisker diagram*.

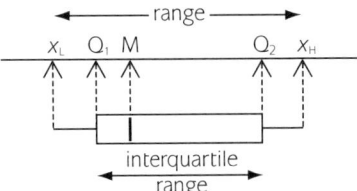

Activity sheet HD9

Looking at data 3

This activity sheet uses a spreadsheet to help you analyse data, produce scatter graphs, look at trends and lines of best fit. Before starting the activities launch Excel.

Open the file transplants.xls containing data on heart and lung transplants and survival rates and select the 'Number of Transplants' tab.

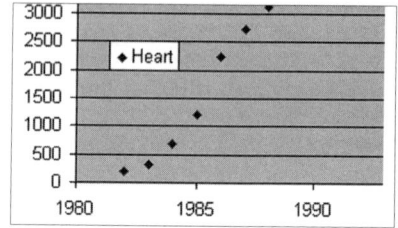

1 **(a)** Highlight the data, in columns A and B down to row 18, showing the number of Heart transplants done from 1982 to 1998

(b) Use the Chart Wizard button on the main toolbar to create a scatter graph of the data. At step 4 of the wizard, place the chart as a new sheet.

(c) Select the 'Number of Transplants' tab again. Highlight the cells from C1, D1 and E1 down to C18, D18 and E18, copy the data, select your 'Chart' tab and paste the data into your scatter diagram and print out a copy of your scatter diagram.

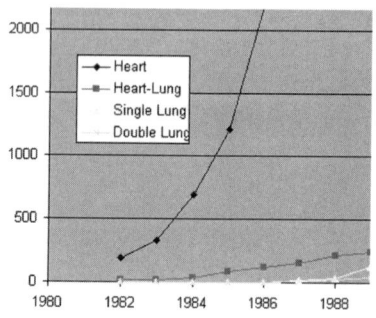

Use your graph to answer the following questions

2 In which year did the number of heart transplants first decrease?

3 Describe the trend of the number of heart transplants from that year.

4 In which year did the number of single lung transplants overtake the number of heart-lung transplants?

5 What is the overall trend of the different types of transplants?

Now select the 'Birth Rates' tab:

6 Write down what the data shows.

7 Create a scatter graph of the data in columns A and B, *using the non-connected points option* and place the chart as a new sheet at step 4 of the wizard.

8 Double click on the *x*-axis and change the Maximum *x* value to 2012 and the Major unit to 1.

(continued)

HD9 – (continued)

9 Use Excel to add a line of best fit and then print out a copy of your scatter graph. ──────────

10 Extend your line of best fit on your printed copy of the graph to estimate the year when the birth rate could fall below 30 for 1000 teenagers.

11 In Excel, on the Birth Rates sheet, highlight all the data from row 2 down to row 11. Use the square at the bottom right of cell B11 to drag down the data until the rate falls below 30.

9	1998	51
10	1999	50
11	2000	49

12 How does this compare with your answer to Question 10?

To produce a line of best fit:
- click on one of your data points in your diagram to highlight all of them then
- click Chart on the menu bar, Add Trendline... and choose the Linear option.

Use this square to drag down the data until the birth rate falls below 30

©David Benjamin and Justin Dodd 2002

Probability models

This activity sheet makes use of a spreadsheet to help you compare experimental probabilities and theoretical probabilities.

1 A normal dice is rolled once.
   (a) Write down, as a fraction, the *theoretical* probability of rolling a 2.
   (b) Use your calculator to change this fraction into a decimal number, accurate to 4 decimal places.
   (c) The dice is rolled 120 times. Write down the *expected* number of 2's.
   (d) The dice is rolled 720 times. Write down the *theoretical* probability of rolling a 2.

Launch Excel and open the file onedice.xls.

2 Click the 'Roll the dice' button once. Compare the experimental probability on the spreadsheet and your answer to Question 1(b), and comment on the difference between the values.

3 Click the 'Roll the dice' button 5 more times. Compare the experimental probability on the spreadsheet and your answer to Question 1(b), and comment on the difference.

4 Experiment to see if the following is true:
'The more times you roll the dice the closer the *experimental* probability will be to the *theoretical* probability.'

Close the file onedice.xls.

5 Two normal dice are rolled together 36 times. In the table below enter:
   ■ The possible totals for the two dice
   ■ The theoretical probabilities of obtaining the different totals.

Totals for the 2 dice											
Probabilities											

Roll the dice

■ Clicking this button in the spreadsheet 'rolls' the dice 120 times.
■ The table calculates the number of times each number on the dice is rolled and calculates the experimental probability for each number.
■ The bar chart shows the results after each 120 rolls of the dice.

Results table	**Line graph**

Click these buttons at any time to see a record of the experimental probabilities for the first 2280 'rolls' of the dice.

(continued)

6 Draw a bar chart below showing the expected number of each total after 144 rolls of the two dice.

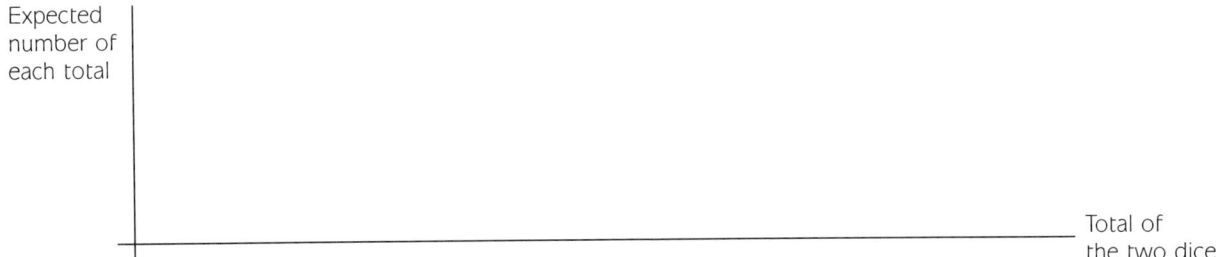

Open the file twodice.xls.

This spreadsheet records the experimental probabilities of rolling two dice.

7 Use the spreadsheet to compare the experimental and theoretical probabilities of rolling two dice.

Close the file twodice.xls and open the file lottery.xls.
This spreadsheet simulates the National Lottery draw.

8 Use this spreadsheet to estimate the number of times consecutive numbers occur in the draw.

Consecutive whole numbers are next to each other:
■ 34 and 35 *are* consecutive
■ 34 and 36 are *not* consecutive.

9 Discuss whether or not it is a good idea to choose consecutive numbers when entering the lottery draw.

Activity sheet HD 11

Looking at data 4

This activity sheet involves using a spreadsheet and includes totals, averages (mean, median and mode), scatter graphs, correlation, and gathering data from the internet. Before starting the activities launch Excel and open the file euro.xls:

Team	Goals Scored	Goals Conceded	Points
Belgium	2	5	3
Czech Republic	3	3	3
Denmark	0	8	0
England	5	6	3
France	7	4	6
Germany	1	5	1
Holland	7	2	9
Italy	6	2	9
Norway	1	1	4
Portugal	7	2	9
Romania	4	4	4
Slovenia	4	5	2
Spain	6	5	6
Sweden	2	4	1
Turkey	3	2	4
Yugoslavia	7	7	4

1 Use the AutoSum button (Σ) :

   (a) in cell B18 to find the total number of goals scored

   (b) in cell C18 to find the total number of goals conceded.

2 **(a)** Write down the modal number of points and name the teams which got that total.

   (b) Enter the formula `=median(D2:D17)` in cell D18 to find the median number of points for the 16 teams.

3 Use a formula:

   (a) in cell B19 to find the mean average number of goals scored

   (b) in cell B20 to find the median average number of goals scored

   (c) in cell B21 to find the modal number of goals scored.

   (d) Comment on the three different averages.

Use these formulae:
(a) =average (B2:B17)
(b) =median (B2:B17)
(c) =mode (B2:B17)

(continued)

4 **(a)** The scatter graph below was created with the Excel chart wizard. Create a similar one displaying the *number of goals scored* plotted against the *number of points gained*. Use the Ctrl key on your keyboard to highlight columns that are not next to each other.

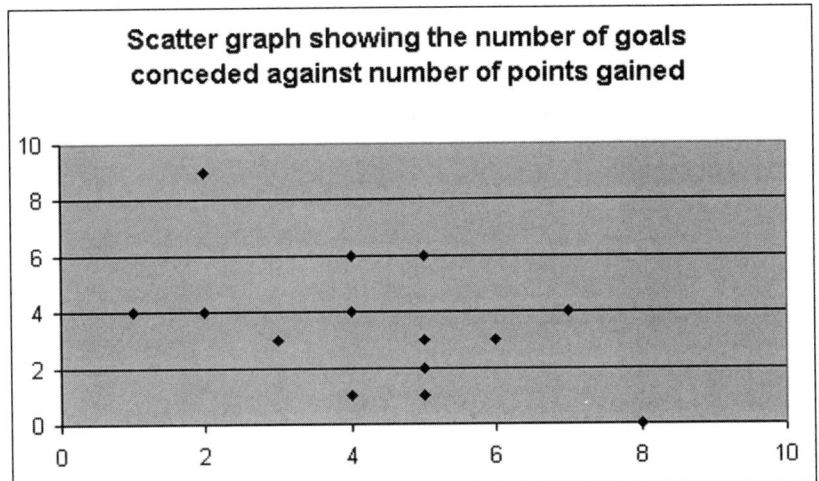

(b) Comment on the two graphs.

5 **(a)** Use the 1999/2000 Carling Premiership final league table to produce two scatter graphs similar to the ones above.

(b) Do the graphs suggest 'more goals means more points' and 'fewer goals means fewer points'?

League tables may be obtained at
http://www.soccerbase.com
and can be copied and pasted directly into Excel.

6 Use the internet site
http://www.teamtalk.com/
and choose your favourite team. Scroll down to the bottom of the page and click on the PLAYERS link. Use the data on the players to make your own statistical analysis.

Estimating the mean

This activity sheet uses a spreadsheet to help you estimate the mean from a group frequency table. Before starting the activities, launch Excel.

1 At a school Spring Fair parents and pupils were asked to estimate the number of marbles in a jar. Enter the data in a spreadsheet shown in the table below.

	A	B	C
	Marbles from	**Marbles to**	**Frequency of guess**
1	**Marbles from**	**Marbles to**	**Frequency of guess**
2	251	300	12
3	301	350	24
4	351	400	88
5	401	450	165
6	451	500	97
7	501	550	74
8	551	600	33
9	601	650	7

(a) Highlight cell C10 and use the AutoSum button to find out the total number of guesses.

the AutoSum button

(b) Enter the label **Midpoint of range** in cell D1 and the formula =(A2+B2)/2 in cell D2. Copy the formula down to cell D9.

(c) Enter the label **Frequency × midpoint** in cell E1 and the formula =C2*D2 in cell E2. Copy the formula down to cell E9.

(d) Highlight cell E10 and use the AutoSum button to find the total of the data in column E.

(e) Use the formula =E10/C10 in cell E11 to find an *estimate* for the mean number of marbles in the jar.

Why is this only an estimate for the mean?

(f) Use the table above to find the interval containing
(i) the median number of marbles guessed
(ii) the modal guess.

(g) Discuss and comment on the three different averages.

(continued)

2 The table below shows the ranges of pulse rates taken from a number of patients at a clinic.

Pulse rate from ...	... to	Number of patients
30	34	1
35	39	3
40	44	7
45	49	24
50	54	45
55	59	62
60	64	79
65	69	83
70	74	92
75	79	70
80	84	42
85	89	31

(a) Use a spreadsheet to estimate the mean pulse rate.

(b) Use the table above to find
    (i) the modal pulse rate
    (ii) the median pulse rate.

(c) Discuss and comment on the three different averages.

Activity sheet HD 13

Investigating standard deviation

This activity sheet makes use of a spreadsheet to help you calculate the standard deviation of sets of data. Before starting the activities, launch Excel.

1 The following data shows the height, to the nearest cm, of 12 garden roses.

58	53	52	47
61	47	64	64
58	44	62	52

(a) Enter the data into cells A2 to A13 of a spreadsheet.

(b) Enter the formula =STDEV(A2:A13) into cell A14 to calculate the standard deviation of the data.

(c) Enter a formula in cell B2 to change the cm into mm and copy the formula down to cell B13. Write down what you think the standard deviation of the changed data should be.

(d) Copy the formula in cell A14 across to cell B14 to check your answer.

(e) In the following month all the roses increased in height by 15 mm. Write down the new standard deviation.

(f) Enter the amended data into column C and check your answer.

2 (a) Open a new spresdsheet. Enter the following data into column A of your spreadsheet and use a formula to calculate the standard deviation of the data.

9	14	8	13
12	15	7	11
13	19	16	15
10	11	6	14
16	12	13	15

(b) Use your answer to write down the standard deviation of data which is twice as much as the given data.

(c) Use column B of your spreadsheet to check your answer.

(d) Write down the standard deviation of the following data

12	17	11	16
15	18	10	14
16	22	19	18
13	14	9	17
19	15	16	18

Every entry in this table is 3 more than the original data.

(continued)

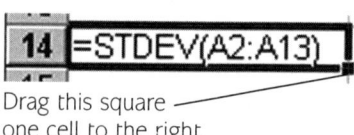

	A
1	**Height in cm**
2	58
3	53

14 =STDEV(A2:A13)

Drag this square one cell to the right

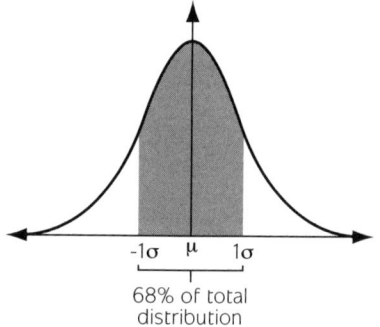

-1σ   μ   1σ

68% of total distribution

Standard deviation, σ, is a measure of the spread of data. For a *normal distribution*, approximately 68% of the data is within 1σ of the mean, μ, and approximately 98% of the data is within 2σ of the mean.

(e) Use column C of your spreadsheet to check your answer.

(f) Write down the standard deviation of the following data

28	43	25	40
37	46	22	34
40	58	49	46
31	34	19	43
49	37	40	46

Every entry in this table was obtained from the original data by multiplying 3 and adding 1.

(g) Use column D of your spreadsheet to check your answer.

Reed Educational & Professional Publishing Limited
Licence and Warranty

This is a legal agreement between you, the end-user, and Reed Educational & Professional Publishing Ltd (REPP) relating to the enclosed software and database product ('the Product'). By installing the Product you agree to these terms. If you do not agree to its terms, return the CD-ROM without installing the Product and all accompanying materials to the place you purchased them and the purchase price will be refunded to you.

1. Copyright
©Reed Educational & Professional Publishing Limited 2002.
A member of the Reed Elsevier plc group of companies. All rights reserved.
The Product and all data, software, documentation and other materials contained therein are the property of REPP or its suppliers and are protected by law.

2. Licence
REPP hereby grants you a non-exclusive, non-transferrable personal licence on the terms in this Agreement to install the Product on a classroom of computers in your school OR to install the Product onto a network provided that the number of users at any one time does not exceed a single class.

3. You may not
a. Copy the Product, except to load it into a computer and to make a single backup copy.
b. Distribute copies of the Product or accompanying user manual to any other person except for use within your institution.
c. Modify, adapt, translate, reverse engineer, decompile, disassemble, or create derivative works based on the Product or user manual.
d. Copy, download, save, store in a retrieval system, publish, transmit, or otherwise reproduce, transfer, save, store, disseminate or use, in any form or by any means, any part of the data contained within the Product.
e. Transfer, resell, or grant any other rights of any other kind to any individual copy of the Product, including the user manual, to any other person.
f. Remove any proprietory notices or labels on the Product or user manual.
g. Allow anyone, including employees, to have access to the Product unless it is necessary to do so in order to utilise it for a purpose authorised by this Agreement.

4. Permitted use
You may copy or print out sections of text or pictures where expressly allowed in the Product and the accompanying materials solely for use within your institution. Apart from this, you may not print off material from the Product.

5. Warranty of original disks
REPP warrants that the original CD on which the Product is delivered is free from defects in materials and workmanship, assuming normal use, for a period of ninety (90) days from the date of purchase. If a defect occurs during this period, you may return the faulty CD to your dealer, along with a dated proof of purchase, and REPP will replace it free of charge. After 90 days you may obtain a replacement by sending your defective CD to your dealer who will charge you a nominal fee for replacement.

6. Please note
Except for the express warranty regarding the original CD in 5 above, REPP gives no other warranty expressed or implied, by statute or otherwise regarding the CD and accompanying materials or their contents, their fitness for purpose, their quality, their merchantability or otherwise.
The liability of REPP under 5 above shall be limited to the amount paid by the customer for the Product. In no event shall REPP be liable for special, consequential or other damages for breach of warranty.
You recognise that the Product is to be used only as a reference aid. It is not intended to be a substitute for the exercise of professional judgement by the end-user.

7. Termination
If you should fail to perform in the manner required in this Agreement, REPP may terminate this Agreement, or exercise any other rights it may have. Upon termination, you shall immediately return all disks containing the Product as well as the user manual. Alternatively, REPP may require that you destroy all of these materials and so certify, in writing, to REPP. All provisions of this licence with regard to the protection of the proprietary rights of REPP shall continue in force after such termination.

8. Applicable law
This Agreement shall be governed by the laws of England and Wales and shall be subject to the non-exclusive jurisdiction of the English courts.